Contributions to Management Science

More information about this series at http://www.springer.com/series/1505

Stéphane Goutte • Khaled Guesmi • Samir Saadi
Editors

Cryptofinance and Mechanisms of Exchange

The Making of Virtual Currency

Springer

Editors
Stéphane Goutte
CEMOTEV, University Paris-Saclay
Versailles, France

Khaled Guesmi
IPAG Business School
Paris, France

Samir Saadi
Telfer School of Management
University of Ottawa
Ottawa, Ontario, Canada

ISSN 1431-1941 ISSN 2197-716X (electronic)
Contributions to Management Science
ISBN 978-3-030-30737-0 ISBN 978-3-030-30738-7 (eBook)
https://doi.org/10.1007/978-3-030-30738-7

This Springer imprint is published by the registered company Springer Nature Switzerland AG.
The registered company address is: Gewerbestrasse 11, 6330 Cham, Switzerland

Preface

The rapid advancement in encryption and network computing gave birth to new tools and products that have influenced the local and global economy alike. One recent and notable example is the emergence of virtual currencies, also known as cryptocurrencies or digital currencies. Virtual currencies, such as Bitcoin, introduced a fundamental transformation that affected the way goods, services, and assets are exchanged. Virtual currencies are experiencing an increasing popularity in the financial markets and in portfolio management as they can be classified as financial asset or commodities on a scale from pure medium of exchange advantages to pure store of value advantages. As a result of its distributed ledgers based on blockchain, cryptocurrencies offer some unique advantages to the economy, investors, and consumers, but also pose considerable risks to users and challenges for regulators when fitting the new technology into the old legal framework. Bitcoin for example may be useful in risk management and ideal for risk-averse investors in anticipation of negative shocks to the market.

Virtual currencies have several properties common to fiat currencies and commodities. Although virtual currencies and their associated technology offer much potential to investors, consumers, businesses, and government entities, they are also the source of risks and challenges to both users and regulators. An important question remains: What does the future hold for virtual currencies? Cryptocurrencies have a dark history of mishaps, especially with the Mt. Gox scandal and the more recent QuadrigaCX scheme. However, the future of cryptocurrencies got brighter with many governments around the world leading the way in the early stages of regulation and taxation. For instance, the Bank of Montreal announced that it will accept dealing with cryptocurrencies as soon as these are regulated and reliable. According to Ben Bernanke, the former Chairman of the Federal Reserve, "[Virtual currencies] may hold long-term promise, particularly if the innovations promote a faster, more secure and more efficient payment system" (Forbes 2014). In fact, the recent implementation of tokenization in Apple Pay and the endorsement of tokenization by Visa show a clear trend for this new practice, which has been at the heart of every cryptocurrency transaction since inception. This fact reinforces the future of cryptocurrencies and shows the advantages of this cryptographic encryption process.

Another potential future of cryptocurrencies is the advent of smart property. Although still at the conceptual level, smart property is likely to change current ownership concepts. Through smart property, an entity can pay electronically to access a cryptographic key giving access to a property temporarily or permanently. Examples encompass any kind of property, whether non-physical (e.g., rights, patents, and trademarks) or physical (e.g., automobile rental and resort timesharing). As an example, in the future world of smart cars, a cryptographic key can ensure that the smart property is secure and no theft can occur. Hence, although cryptocurrencies are very volatile and risk extinction, the technology behind them is here to stay and may become a major player in the future smart world. Some contend that the chances of an evolved form of cryptocurrency will replace paper money by 2050 if that form is backed by a solid commodity or a set of commodities. The adoption of the technology behind the cryptocurrencies has the potential to validate any single transaction every person makes on the planet.

Money may be likely to be treated as bits of information in the future. If and when that happens, money exchange will be similar to information exchange. The next generation of developers will create cryptographic breakthroughs that revolutionize financial transactions. Although the current concept of cryptocurrency will evolve over time, the technology and innovation that come with it will open the way to something much bigger—a revolution not only in financial transactions but also in every transaction.

This book provides a comprehensive discussion on the important issues related to cryptocurrencies ranging from pricing, financial, and legal to technological aspects.

Versailles, France — Stéphane Goutte
Ottawa, ON — Khaled Guesmi
Paris, France — Samir Saadi

Contents

Cryptocurrencies as an Asset Class 1
Sinan Krückeberg and Peter Scholz

Are Virtual Currencies Virtuous? Ethical and Environmental Issues . . . 29
Sondes Mbarek, Donia Trabelsi, and Michel Berne

Cryptocurrency Mining 51
Vikrant Gandotra, François-Éric Racicot, and Alireza Rahimzadeh

Regulating Bitcoin: A Tax Case Study 69
Margaret Ryznar

Are Cryptocurrencies Truly Trustless? 77
Usman W. Chohan

Blockchain and Alternative Sources of Financing 91
Othalia Doe-Bruce

Tokenomics 113
Ralf Wandmacher

Crypto Tokens and Token Offerings: An Introduction 125
Chen Liu and Haoquan Wang

Initial Coin Offerings: What Do We Know and What Are the Success Factors? 145
Chen Liu and Haoquan Wang

Initial Coin Offerings (ICOs): Risks, Regulation, and Accountability . . . 165
Usman W. Chohan

Cryptocurrencies and Risk Mitigation 179
Haifa Amairi, Boushra El Haj Hassan, and Ahlem Zantour

Index 193

About the Editors

Stéphane Goutte has two PhDs, one in Mathematics and one in Economics. He received his Habilitation for Supervising Scientific Research (HDR) in 2017 at University Paris Dauphine. He is a full Professor at CEMOTEV, Université Versailles St.-Quentin-en-Yvelines, France. He teaches mathematics and related topics in MSc and BSc. He is also a Senior Editor of *Finance Research Letters*; an Associate Editor of *International Review of Financial Analysis* (IRFA) and *Research in International Business and Finance*; a Subject Editor of *Journal of International Financial Markets, Institutions and Money* (JIFMIM); and an Editorial member of *European Management Review* (EMR). His interests lie in the area of mathematical finance and econometric applied in energy or commodities. He has published more than 40 research papers in international review. He has also been a Guest Editor of various special issues of international peer-reviewed journals and Editor of many handbooks.

Khaled Guesmi is a full Professor of Finance at Paris School of Business and Adjunct Professor at Telfer School of Management, University of Canada. He undertakes research and lectures on empirical finance, applied time-series econometrics, and commodity markets. Dr. Guesmi obtained his HDR (Habilitation for Supervising Doctoral Research) in July 2015. He holds his PhD in Economics from the University Paris Nanterre in 2011, and his MSc in Finance from Paris I University of Sorbonne in 2005. Previously, he served as Professor of Finance and Head of Environment, Climate Change and Energy Transition Chair at IPAG Business School, as Associate Research positions at "EconomiX" laboratory at the University of Paris Ouest La Défense and "ERF" Economic Research Forum, Egypt. In 2003, Dr. Guesmi joined the UNESCO as a Research Manager, and in 2008, he joined "*Caisse de Dépôts et Consignations*" as Financial Analyst.

He has published articles in leading refereed journals, including the *Energy Journal*; *International Review of Financial Analysis; Journal of International Financial Markets, Institutions & Money, Annals of Operations Research, Energy Economics and Energy Policy*. Furthermore, Dr. Guesmi currently serves as

Associate Editor at *Finance Research Letters*, as an international advisory board at *The International Spectator*. In addition, Dr. Guesmi is the founder of the International Symposium on Energy and Finance Issues and the Project Manager of the European Commission's Horizon 2020 Program for Research and Innovation. Dr. Khaled Guesmi holds the 4th best world researcher in Energy Finance: Frontiers and Future Development in Beijing in 2018.

Samir Saadi is an Associate Professor and Telfer Fellow in Behavioral Finance at Telfer School of Business. He received a PhD from Smith School of Business of Queen's University. Before joining the University of Ottawa, he was a Visiting Scholar at world leading research-intensive institutions such as Stern School of Business of New York University and INSEAD (France). He held several external grants from, among others, the Social Sciences and Humanities Research Council of Canada (SSHRC). His current research funded by SSHRC focuses on how social media influences investors' behavior. Along with several book chapters, Dr. Saadi has published over 50 academic papers in refereed journals such as *Financial Management*, *Journal of Corporate Finance*, *Journal of Banking and Finance*, *Journal of Financial Research*, *Contemporary Accounting Research*, *Journal of Business Ethics*, and *Journal of Business Finance and Accounting*. He is an Editor of *Finance Research Letters* and has been a Guest Editor of several special issues of peer-reviewed journals, and keynote speaker in several international conferences. He is the recipient of best paper awards, and various prestigious awards from, among others, the SSHRC, the Europlace Institute of Finance, and the American Finance Association.

Cryptocurrencies as an Asset Class

Sinan Krückeberg and Peter Scholz

Abstract Cryptocurrencies are a new emergence at the intersection of technology and finance. It is therefore of particular interest whether cryptocurrencies can form a new asset class or need to be subsumed under an existing one. We find that cryptocurrencies show characteristics of a distinct asset class based on strong internal correlation, an absence of correlation with any traditional asset class as well as sufficient market liquidity, while market stability has room for improvement. Adding cryptocurrency to traditional portfolio structures may lead to significant and persistent risk-adjusted outperformance. These results support the careful introduction of cryptocurrencies into the asset management mainstream.

1 Introduction

Economic turbulences such as the Subprime Crisis have served to highlight the fragility of our monetary and financial system. As a reaction to what has become one of the most severe crises in history, *Bitcoin* was launched in 2009 setting the stage for a multitude of further projects which led to the development of a new bridge between technology and finance: *cryptocurrency*. Increasing investment continues to flow into the sector amounting to a total market capitalization of over US$400 billion by December 2017. However, it remains an essential question whether cryptocurrencies can qualify as a distinct asset class in their own right, enabling diversification and outperformance compared to portfolios comprising only traditional asset classes. If cryptocurrencies were to constitute a distinct asset class, this would carry significant implications for fund managers, regulators and policy makers alike.

S. Krückeberg (✉)
Krückeberg Family Office, Hamburg, Germany
e-mail: sinan@krueckeberg.us

P. Scholz
Hamburg School of Business Administration (HSBA) Department of Banking & Finance, Hamburg, Germany
e-mail: peter.scholz@hsba.de

S. Goutte et al. (eds.), *Cryptofinance and Mechanisms of Exchange*, Contributions to Management Science, https://doi.org/10.1007/978-3-030-30738-7_1

While existing literature has touched upon the nature and performance impact of cryptocurrencies, significant scope remains to form a comprehensive picture. Attempts at answering the question of whether cryptocurrencies are investable and constitute a distinct asset class have so far exclusively focused on correlation between Bitcoin, as a proxy for cryptocurrencies, and traditional assets. We increase the granularity of inquiry by covering a broad selection of individual cryptocurrencies and extend as well as embed correlation analyses in a theoretical framework for the definition of asset classes provided by Sharpe (1992). Beyond correlation, we add liquidity and stability as criteria to evaluate the investability of cryptocurrency. On the side of analyses regarding the impact of cryptocurrency on portfolio performance, we significantly extend time-series length and provide multiple weighting methods that aim to reflect implementable allocations to cryptocurrency and thereby portfolio structures that can be applied in asset management practice. We thereby aim to provide a robust fundament to comprehensively evaluate cryptocurrencies as an asset class as well as their real-world impact on portfolio performance.

Under the umbrella term cryptocurrencies, we therefore differentiate between cryptographic *coins*, which use their own blockchains, and *tokens*, which operate atop a third party's blockchain architecture. The 10 largest coins and tokens by market capitalization as of mid-December 2017 with at least a price history of 3 months are selected. First, we are interested whether either coins or tokens or both can qualify as asset classes in their own right. We evaluate cryptocurrencies on the basis of parametric and non-parametric correlation measures, market liquidity and market stability against Market Wide Circuit Breaks and Limit Up-Limit Down triggers. Second, we are interested whether adding cryptocurrencies to traditional portfolios will lead to superior results regarding the Sharpe ratio for quarterly rebalancing intervals via *ex-post* optimizations. Third, we use the results of previous *ex-post* optimizations for *ex-ante* portfolio calibration. Three different weighting approaches are applied. *Dynamic* weighting uses the dynamic quarter-by-quarter allocations of the *ex-post* optimization to rebalance portfolios. *Average* weighting employs the average weights for the respective asset classes over all optimized quarters with rebalancing to initial weights at the end of each quarter. *Conservative* weighting utilizes traditionally defensive portfolio allocations vs. such allocations plus the addition of 1% of the asset class cryptocurrencies. Thereby, we test for risk-adjusted outperformance of portfolios containing cryptocurrencies versus portfolios that only contain traditional asset classes via three different weighting rules.

We find that cryptocurrencies qualify as a distinct asset class. Strong correlation among cryptocurrencies is contrasted by almost no statistically significant correlation of cryptocurrencies with traditional asset classes. Absolute market liquidity for some cryptocurrencies is already on equal footing with traditional equities while liquidity in relation to market capitalization is significantly stronger for cryptocurrencies than for traditional assets. Market stability needs to improve as evidenced by numerous market breaks as well as Limit-Up-Limit-Down trigger signals. Moreover, we find evidence for the existence of two distinct sub-asset classes coins and tokens. *Ex-post* portfolio optimizations employing the Sharpe ratio show that adding cryptocurrencies to traditional portfolios leads to superior results regarding risk-adjusted returns. Using the previously calculated *ex-post* Sharpe-optimal weights for *ex-ante* portfolio calibration, dynamic weighting underperforms while both the averages and conservative weightings consistently outperform.

These findings imply that investment practitioners can find attractive upside and diversification effects in adding even small cryptocurrency positions to their portfolios. Furthermore, defining cryptocurrencies as an asset class could have an impact both on regulatory treatment of such as well as future policy debate.

2 The Technology Behind Cryptocurrency

Merkle's (1980) seminal work has tied together essential strands of research on protocols for public key crypto systems, forming a vital foundation for the future development of cryptocurrency. Merkle reviews both conventional and digital cryptographic protocols and concludes that centralized key distribution for some use cases is inferior to public key distribution, due to vulnerabilities regarding loss of security and function as well as proneness to destruction (for types and merits of decentralization, see also Buterin 2017b). He provides the key building blocks for future development of decentralized cryptocurrencies by outlining Authenticated Public Key Distribution, a Basic Digital Signature Protocol, Time Stamping and Witnessed Digital Signatures. Authenticated Public Key Distribution establishes a system in which each participant holds a randomly computed public enciphering key as well as a private deciphering key. Encrypted information is signed with the sender's private key and encrypted with the recipients public key. This way, information transmitted can be authenticated as sent by the sender while only being decipherable by the recipient. To implement a full-fledged cryptocurrency, a consensus mechanism, generating consensus about the legitimate state of a system, is needed in addition to a general encryption mechanism. Such consensus mechanisms are used to record valid transactions by implementing Time Stamping and Witnessed Digital Signatures. Time stamps provide a proof of existence for each transaction at or before a certain point in time, while Witnessed Digital Signatures serve as proof of validity. The combination of an encryption protocol together with a consensus mechanism enables the maintenance of a public ledger of transactions. The technology for such a ledger is most prominently realized as a *Blockchain* (for a primer on Blockchain technology, see Voshmgir and Kalinov 2017).[1] Individual transactions are aggregated into blocks by individual participants of the system for a reward and subsequently integrated into a chain of blocks and marked with a time stamp. Individual blocks in a chain subsequently get confirmed by other participants of the system through the addition future blocks atop the previous block. In the case

[1]Bleeding edge alternatives such as the 'Tangle' and 'Hashgraph' are emerging, see Popov (2017) and Baird (2016) respectively.

of branching of the blockchain, one branch will ultimately 'outgrow' others, emerging as the dominant branch while the alternative branches will 'die off'. A copy of the decentralized ledger of transactions is held by each participant of the ecosystem. In order to allow for scaling of the public ledger, Merkle Trees are utilized to minimize storage needs (Merkle 1990). Initial attempts seeking to implement such a design can be found in Wei Dai's *B-Money* (WeiDai 1998) followed by Szabo's *BitGold* (Szabo 2008). *Bitcoin* has so far without doubt emerged as the most prominent system based on Nakamoto's work (Nakamoto 2008). Buterin (2013a, b) proposes *Ethereum*, launching a platform with increased functionality (see also Wood 2014; Buterin 2016). Upon Ethereum, a large variety of projects has been built, such as *EOS*, *FileCoin* and *Golem* (cf. EOSProject 2017; ProtocolLabs 2017; GolemProject 2016).

Three mechanisms for achieving consensus regarding the validity of a mined block are Proof of Work (PoW), Proof of Stake (PoS) and Proof of Burn (PoB) which will be explained below. These mechanisms are primarily relevant for reasons of system security, aiming to make the counterfeiting of the distributed ledger as expensive as possible to prevent attacks on ledger integrity. However, tightly linked are economic implications regarding token supply as well as volatility arising from possible insecurity. Together, these factors have given rise to the emerging field of *Cryptoeconomics*, seeking to balance considerations of cryptography and economic incentives.

In a PoW mechanism, the influence individual miners can exert on the development of the blockchain is defined by the computational effort or work invested into the maintenance of the system. The work invested is directed at solving a computational puzzle as originally described by Dwork and Naor (1992), who propose the implementation of pricing functions in order to gain access to certain information. Jakobsson and Juels (1999) formalize the concept of PoW. The meaning is hereby shifted from an authentication mechanism to a verification of computational resources invested during a certain period of time. Juels and Brainard (1999) highlight PoW schemes as protection against the flooding of a server with requests to carry out denial-of-service attacks. This development has culminated in Back's (2002) *Hashcash* cost function, which currently forms the basis for most crypto tokens in circulation. Miners willing to mine an incremental block of transactions are provided with a randomly generated hash. The aim of the miner is then to iterate a nonce a vast amount of times until one is found that conforms to the required number of zero bits on the resulting bit string. The number of zero bits required defines the difficulty of the puzzle and therefore the frequency at which blocks will be mined. The nonce as the solution to the puzzle is difficult to compute yet simple to verify, which allows all participants to easily check and consensually confirm the validity of a nonce, ending the puzzle and generating a new hash. Once computing power has been invested into finding an appropriate nonce, the newly mined block can not be changed without redoing the entire work, not only of the last incremental block but of all following blocks, since all blocks are linked to each other transitively with the longest chain commanding legitimacy. This provides increasing protection against double-spending with increasing length of a Blockchain as detailed by Rosenfeld (2012).

Proof of Stake consensus is based on the expectation that token holders of a certain crypto system are interested in successfully maintaining an accurate ledger of transactions. The participant with the largest relative stake in the crypto asset is determined as the miner of the incremental block. There exists no mining reward, thus token supply need not be inflated. However, transaction fees can be collected by the miner to incentivize participation. This approach is less resource intensive while security is assured through the self-interest of participants not to implement malicious transactions and therefore protect the value of participants' own token holdings. An extension of this concept is Delegated Proof of Stake, through which stakeholders 'elect' the miners of an incremental block (cf. EOSProject 2017).

Within Proof of Burn systems, miners "bid" for the right to mine an incremental block by sending existing tokens to a burn address (for an example, see P4Titan 2014). This burn address is predetermined and invalid, tokens sent to it get "burned", that is, they disappear and the token supply decreases by that specific amount. This leads to a relative wealth transfer to all other token holders since the existing value of the system is now divided by a token quantity which is smaller by exactly the amount of tokens burned. The participant sending the largest amount of tokens to the burn address has the right to mine the incremental block and to collect transaction fees. Each miner will bid exactly that amount of tokens for which he can still make a profit after accounting for tokens burned and equipment as well as opportunity costs invested. Proof of Burn can be validated by back checking transfers to the burn address. In contrast to a PoW mechanism, resources invested do not take the form of mining equipment but rather tokens burned. Again, security is ensured by the self-interest of participants and comparative cost for attackers that need to be incurred to break the system.

Consensus mechanisms can generally be used in parallel and switches between mechanisms can occur. Tying all considerations above together, the combination of an encryption protocol with a consensus mechanism then enables the existence of a Decentralized Autonomous Organization (DAO). Such a DAO constitutes a network for the internal transaction of value that is governed by the automatic mechanisms of a blockchain. Adding a cryptographic currency that can be traded between participants then finally enables the operation of Decentralized Applications (DApps) that provide value to participants. It is these currencies specifically, which will be evaluated as to their suitability to constitute a distinct asset class, leading to diversification effects and potential outperformance.

3 The Rise of Cryptocurrency

The starting point for the rapid development of cryptocurrency is marked without doubt by the publishing of the Bitcoin whitepaper in October of 2008, incepting the cryptocurrency that has ever since evolved to become the sector stalwart. The first transaction of Bitcoin (BTC) was soon thereafter executed between the Bitcoin founder(s) 'Satoshi Nakamoto' and cryptographer Hal Finney in 2009, totalling 10 BTC (Higgins 2014). The cryptocurrency sphere as a whole subsequently started to broaden in 2011 with the introduction of Litecoin, among others, as a variation of Bitcoin. 2013 brought the first instance of a so called fork of the Bitcoin network, albeit accidental—an event describing significant changes to the currency's underlying protocol splitting the block chain into two competing strands. The issue was quickly resolved, however, marking a significant step forward in Bitcoin's maturation process (Buterin 2013a). Regarding the surrounding infrastructure servicing cryptocurrencies, 2014 saw the most significant cryptocurrency exchange Mt. Gox file for insolvency after a significant amount of Bitcoin had been stolen (Dougherty and Huang 2014). Nevertheless, Bitcoin and by extension the crypto sphere as a whole recovered anew, underlining cryptocurrencies' resilience against external shocks. 2016 brought a crucial step forward in the technology of cryptocurrencies with the launch of the Ethereum network, henceforth enabling the implementation of DApps. Subsequently, in the governmental domain the Swiss canton Zug introduced the initiative 'CryptoValley' aiming to become a global hub for the cryptocurrency sector and allowing fees owed to the government to be paid in Bitcoin up to the amount of 200 CHF (Swissinfo 2016).

The year 2017 saw a further Swiss initiative, this time by the city of Chiasso in the canton Ticino, allowing tax payments in Bitcoin up to the amount of 250 CHF, tied to the initiative 'CryptoPolis' (Allen 2017). Other countries, such as Estonia, have launched even more comprehensive programs which, however, still await final implementation (Ummelas 2018). Crucially, in the United States both the Chicago Board Options Exchange (10th of December 2017) and Chicago Mercantile Exchange (17th of December 2017) launched Bitcoin futures trading in short sequence, paving the way for Bitcoin to access established capital markets infrastructure for the very first time (cf. CBOE 2017; CME 2017). With cryptocurrency gaining more and more legitimacy within the wider public realm, institutional arrangements, regulatory frameworks and infrastructure continue to evolve and adapt. With cryptocurrencies having risen from an initial total market capitalization of US$1.3 billion in early 2013 to a peak of US$813 billion as of early 2018, special interest lies in the question whether cryptocurrency can qualify as a new asset class and if so, whether adding this asset class can generate outperformance against traditional portfolios.

4 Potential Assets: Cryptographic Coins and Tokens

When evaluating the suitability and performance of cryptographic currencies as assets, different types of DApps can be distinguished along multiple criteria, leading to a classification of different cryptocurrencies. Since cryptocurrencies operate within the framework of a DApp, the nature of the DApp is important to derive the value basis for the currency. Following Johnston et al. (2015), three types of

DApps can be distinguished. Type I operates its proprietary blockchain, protocol and currency. Type II uses its own protocol and currency but not its own blockchain and therefore operates on the blockchain of a Type I DApp. Type III is a protocol that uses its own currency, however based on a protocol of a Type II DApp and the blockchain of a Type I DApp.

In order to simplify further economic analysis, we will distinguish between cryptographic *coins* that are used in DApps of Type I versus cryptographic *tokens* that run on Types II and III. The reason for this distinction is that coins are native units of an independent system and often find primary application in functioning as a means of payment, while tokens are non-native units usually securitizing additional utility. Therefore tokens are sometimes also referred to as *utility tokens*. This classification seems akin to the classic monetary theory's fiat money and commodity money, however, the lines seem to be too blurred for such clear cut distinction.

Ultimately, the value of a DApp maintained by the DAO is 'securitized' by coins and tokens (for approaches to token valuation, see Buterin 2017a; Kalla 2017). Empirical analyses regarding the question whether cryptocurrencies can constitute an asset class in their own right, therefore need to focus on the empirical properties of coins and tokens.

5 Cryptocurrencies as Investments

Contributions on cryptocurrencies as investments have to date followed two strands of analysis. One strand focuses on the question of investability, specifically an absence of correlation between cryptocurrencies and traditional asset classes, to set cryptocurrency apart as a distinct asset class. A second strand focusses on the potential performance impact that cryptocurrency can have when added to traditional portfolios.

5.1 Investability

Multiple contributions attempt to shed light onto the investability of cryptocurrency, all of which focus on correlation as the measure to distinguish cryptocurrencies from traditional asset classes. Briere et al. (2015) study weekly return data for Bitcoin and a broad range of traditional asset class indices for developed and emerging economies from July 2010 to November 2013. They find that correlation between Bitcoin and traditional asset classes is negligible and therefore conclude that cryptocurrency seems to form an attractive new investment opportunity. Eisl et al. (2015) concur, having studied correlations between Bitcoin returns and a range of traditional asset class index returns from July 2010 to April 2015. Lee et al. (2018) follow a similar approach in studying correlations between the cryptocurrency index CRIX and various indices of traditional asset classes from August 2014 to March 2017, confirming previous findings of low correlation and therefore cryptocurrency as a new investment opportunity.

However, we identify multiple gaps in the data coverage and methodology of existing studies which provide opportunity to extend and complement the body of literature. Due to the rather short history and limited data availability of cryptocurrency at the time of writing, early contributions such as by Briere et al. (2015) are limited in both length and granularity of the time series by only studying weekly closing prices from 2010 to 2013. Despite extending time series length by one and a half years to April 2015, Eisl et al. (2015) still remain limited in sample length. Unfortunately, Eisl et al. do not to explicitly specify the granularity of data used. Lee et al. (2018) use a comparatively long dataset ending in March 2017. However, all data sets used in previous studies are either directly focused on Bitcoin as a proxy for cryptocurrency or dominated by Bitcoin via its disproportionate representation in the CRIX index. The CRIX index, while providing an improvement in coverage of cryptocurrencies, blends cryptocurrency prices based on market capitalization weighting and thereby smooths individual fluctuations of cryptocurrencies by supplying one aggregate figure. While compromises will need to be made regarding the quantity of cryptocurrencies analyzed, we see potential for even clearer insight into the nature of cryptocurrency by analyzing individual cryptocurrencies. Besides correlation between cryptocurrencies and traditional asset classes, a further open question remains whether correlation within the group of cryptocurrencies is significant enough in order to constitute one single asset class. This question is of particular interest since, as we show above, cryptocurrency can naturally be subdivided into coins and tokens. Both an analysis of intra-cryptocurrency correlation and differences along the coin/token distinction have not been supplied to date.

Correlation analyses on their own; however, appear to paint an incomplete picture when aiming to answer the question of investability and the potential for cryptocurrency to form a distinct asset class. Nevertheless, other factors playing a role for the investability of cryptocurrency have so far received little attention. Some studies have acknowledged liquidity as an important factor for the ability to enter and exit investment positions. Fink and Johann (2014) observe that the price impact of individual trades decreases from 2011 to May 2014 while absolute liquidity increases over the same timeframe. Briere et al. (2015) consider liquidity in passing, mentioning a general increase of absolute liquidity over time without providing a specific measure or benchmark. Dyhrberg et al. (2018) briefly touch upon the subject of absolute Bitcoin liquidity which is, however, limited to three exchanges and does not put primary focus on such analysis. Trimborn et al. (2018) compare average absolute liquidity across 42 cryptocurrencies to that of the S&P500 and find that such averages are lower for cryptocurrencies than for the S&P500. Wei (2018) studies the impact of liquidity on return predictability and volatility, as well as an illiquidity premium. She finds that both return predictability and volatility decrease as liquidity increases, while no evidence could be found of an illiquidity premium. Elendner et al. (2018) provide an overview of absolute daily liquidity from April 2014 to June 2016. While, therefore, liquidity has been acknowledged as a factor for investability, we find scope to extend previous contributions by studying individual cryptocurrencies rather than Bitcoin alone, by distinguishing between the two groups of coins and tokens, as well as by studying relative liquidity.

Besides correlation and liquidity, the factor stability has not received attention so far. However, besides having defined cryptocurrency as a distinct asset class via correlation and knowing that adequate liquidity is available to enter and exit investment positions, market stability seems an essential feature for investability, as market breaks due to high volatility might lead to frequent halts in trading that can negatively impact investability of cryptocurrency.

5.2 *Impact on Portfolio Performance*

Within the second strand of literature turning to the impact of cryptocurrency on portfolio performance, to the best of our knowledge, the first contribution seems to be Briere et al. (2015). By using a time series of weekly Bitcoin data over approximately three and a half years (July 2010 to Dec 2013), the performance of different portfolios consisting of traditional as well as alternative asset classes is explored. As weighting-schemes, Briere et al. (2015) use equal-weighted portfolios and Markowitz mean-variance optimization. For both weighting schemes, portfolio performance including and excluding Bitcoin is analyzed over the three and a half years without rebalancing. Due to the time series properties of Bitcoin within this timeframe—high returns, high volatility but low correlation—this leads to superior performance as measured by the Sharpe ratio for the portfolios including Bitcoin in both weighting schemes.

Eisl et al. (2015) extend the approach of Briere et al. (2015) by applying the CVaR (Conditional Value-at-Risk) instead of variance as a measure of risk. Furthermore, the "single point in time" approach is replaced by a backtest with rebalancing as well as the introduction of different constraints. Despite the extended framework, results widely confirm the findings of Briere et al. (2015). Inclusion of Bitcoin with a weight between 1.65% and 7.69% appears valuable "even in already well-diversified portfolios".

A further contribution which analyzes the performance of portfolios including cryptocurrencies as an asset has been supplied by Lee et al. (2018). Using the cryptocurrency index CRIX and a time series of approximately two and a half years, Lee et al. (2018) explore the performance of a portfolio consisting of similar assets like those of Eisl et al. (2015) while following the CVaR and mean-variance approach and comparing the performance of both optimization methods. However, they also rely on a "single point in time" approach like Briere et al. (2015). In general, the results of Briere et al. (2015) and Eisl et al. (2015) are confirmed, in that an inclusion of cryptocurrencies improves the risk/return characteristics, especially for the minimum-variance portfolio. Interestingly, Lee et al.'s weights for the CRIX of up to 72.5% seem to be significantly higher than those of Eisl et al. (2015).

We add to the existing literature in multiple ways. Our time series stretch more than 4 years of daily data while we also employ different rebalancing schemes applied on a quarterly basis and go beyond ex-post optimization. While Eisl et al.

(2015) apply out-of-sample tests by using the optimized weights of each previous quarter for rebalancing, we supply additional insight by using three different methods. First, dynamic weighting quarter-by quarter; second, average weighting across all quarters with rebalancing to initial weights at the end of each quarter; and third, a static allocating of 1% to Bitcoin across each quarter. By and large, we use similar traditional asset classes for diversification as employed by Eisl et al. (2015) by including stocks, bonds, real estate, gold, and oil. In contrast to previous findings, however, we do not detect beneficial impact from adding Bitcoin to minimum-variance portfolios. More importantly, we do find general benefit from adding Bitcoin to traditional portfolios. An allocation of Bitcoin to traditional portfolio structures as low as 1% significantly increases the Sharpe ratio. However, due to high volatility and negative returns on a quarterly basis, assigning large allocations to cryptocurrency can have significant negative impact on portfolio performance.

The following chapters will present our analyses that aim to extend existing literature discussed above in order to close remaining research gaps and supply a sound fundament on which to evaluate whether cryptocurrency does in fact constitute a distinct asset class and can lead to improvements of portfolio performance when added to traditional portfolios.

6 Dataset and Methodology

6.1 Data

We use the platform *coinmarketcap.com* as our data source for correlation and liquidity analyses, which grants open access to their data for any use or purpose. Coinmarketcap.com aggregates the daily volume weighted average prices and the total trading volume for more than 1300 cryptocurrencies over all cryptocurrency exchanges that the respective currencies are listed on. The data consist of daily opening, high, low and closing prices as well as trade volume and market cap time series. All data relate to the 24 h window of UTC—Coordinated Universal Time (for details, see separate annex). All data have been sourced by December 8th, 2017. We select the top 10 coins and tokens by market cap, respectively, as of December 8th, 2017, 11:00UTC, for cryptocurrencies with a price history >3 months (Fig. 1).

For analyses regarding market stability, we download tick-by-tick data from *Poloniex*, one of the largest cryptocurrency trading platforms, through its native API (for details, see separate annex). Poloniex supplies quotes against the Tether (USDT), a dollar-pegged cryptocurrency which is employed as a dollar surrogate on major exchanges. While the Tether shows episodes of fluctuation around the perfect peg of USD\USDT 1, prices quoted in USDT are well within the individual range of prices in USD as quoted on different globally operating exchanges. Nevertheless, we source daily USDT closing prices against the USD from the platform coinmarketcap.com to eliminate potential noise added to daily tick-by-tick data by the USDT.

Rank	Coin	Market Cap	Token	Market Cap
1	Bitcoin (BTC)	US$ 265,539,107,455	EOS (EOS)	US$ 2,349,608,345
2	Ether (ETH)	US$ 43,175,564,793	Populous (PPT)	US$ 1,065,186,620
3	Bitcoin Cash (BCH)	US$ 25,196,957,992	OmiseGO (OMG)	US$ 837,498,937
4	IOTA (MIOTA)	US$ 11,761,165,463	Ardor (ARDR)	US$ 518,629,589
5	Ripple (XRP)	US$ 9,626,871,230	Veritaseum (VERI)	US$ 311,322,142
6	Litecoin (LTC)	US$ 5,451,485,929	Augur (REP)	US$ 294,514,000
7	Dash (DASH)	US$ 5,397,079,086	Golem (GNT)	US$ 236,430,309
8	Monero (XMR)	US$ 4,132,314,200	Binance Coin (BNB)	US$ 232,213,574
9	Ethereum Classic (ETC)	US$ 2,639,390,342	MaidSafeCoin (MAID)	US$ 215,912,303
10	Stellar Lumens (XLM)	US$ 2,394,395,573	TenX (PAY)	US$ 207,612,454

Fig. 1 Top 10 coins and tokens by market cap with price history >3 months as of the 8th of Dec. 2017 have been included

Indices	Equities	Currencies	Bonds	Commodities	Real Estate
S&P500	Facebook (A)	EUR\USD	J.P. Morgan	WTI Spot	MSCI World
NASDAQ100	Amazon	USD\CHF	Government Bond	NYMEX Brent Crude	Real Estate
FTSE100	Apple	USD\JPY	Index	Gold Handy & Harman	Price Index
EUROSTOXX50	Netflix	USD\SGD			
DAX30	Alphabet (A)				
TecDAX30	Alibaba ADR				
Hang Seng	Baidu ADR				
Nikkei225	RenRen ADR				
	Tencent ADR				

Fig. 2 All traditional asset classes included in empirical analyses

Financial market data regarding traditional asset classes are sourced from Reuters Datastream and Bloomberg Terminal for the time period from April 28th 2013 to November 3rd 2017 (Fig. 2). Data include daily open, close, high and low prices, daily volume and market cap for relevant assets on the respective exchanges. The selection of traditional asset classes and particular assets within a class has been guided by multiple considerations. With the inclusion of equities, bonds, currencies, real estate and commodities it is the aim to incorporate all essential classes typically available to a reasonably sophisticated investor. Among indices, we select those that represent the globally most significant exchanges and simultaneously cover a sufficiently broad geographic area. In addition to equity indices, we include specific equities from the tech sector for purposes of comparability with cryptocurrency as a phenomenon emerging from the sphere of technology. Special consideration will be given to the FAANG group of stocks (Facebook, Amazon, Apple, Netflix, Google/Alphabet) due to their significance to the sector. Regarding currencies, we focus on the four globally most relevant pairs while the individual assets from within the remaining three classes (Bonds, Commodities and Real Estate) are selected considering relevance and geographic scope.

Portfolio optimization simulations rely on data as reported above.

6.2 Methodology

Our methodology is guided by three aims. First, to analyze cryptocurrencies' suitability to form a distinct asset class. For the definition of an asset class, we follow Sharpe (1992) who distinguishes asset classes along the three criteria of (1) mutual exclusivity between asset classes, (2) exhaustiveness within an asset class and (3) differing returns between asset classes. Preceding the analysis of mutual exclusivity *between* asset classes we test for mutual necessity, that is, the correlation *within* the class of cryptocurrency in order to be able to define cryptocurrency as a consistent whole. Internal exhaustiveness is reached by considering the entire spectrum of cryptocurrency and selecting the currencies representing the lion's share of total market capitalization. The differing returns criterion as well as mutual exclusivity are satisfied via correlation analyses between the group cryptocurrency and all traditional asset classes. To the necessary condition of correlation, we add the sufficient conditions liquidity and stability. The liquidity analysis is aimed at comparing liquidity of crypto markets relative to traditional tech equities as an indicator whether investment positions can effectively be entered and exited. The stability criterion serves as a test for the maturity of the cryptocurrency space as a whole. Second, we analyze whether there are significant differences between cryptographic coins and tokens, requiring the creation of sub-groups within the potential new asset class. Third, we evaluate whether portfolio structures can either increase returns or mitigate volatility by including cryptocurrencies. This is done both via quarterly *ex-post* optimization of portfolio Sharpe ratios, as well as *ex-ante* portfolio construction using three different structuring approaches.

6.2.1 Correlation

Inputs into the correlation analyses are simple daily returns. We test all time series for normality using the Kolmogorov-Smirnov test with Lilliefors and Stephens modification as well as by estimating the Shapiro-Wilk *W*. Due to non-normality of the cryptocurrency time series, correlation is estimated via three different measures: the parametric Pearson's *r*, as well as the non-parametric Kendall's tau and Spearman's rho. Parametric and non-parametric results serve as a robustness check for each other while both non-parametric tests serve as an internal consistency check. We test for dependencies between individual cryptocurrencies as well as between cryptocurrencies and traditional asset classes. Thereafter, coins and tokens are analyzed for correlation between these two sub groups using equally weighted mean and median daily returns of the groups. All time series are tested for time dependent variations. To gain deeper insight we further analyze the correlation of daily returns of an individual title with the correlation between this specific title and all other titles in the respective group. This is done both for all cryptocurrency pairs as well as among all FAANG stock pairs. We thereby aim at conclusions regarding the question whether cryptocurrencies and traditional FAANG stocks show comparable return dependent correlation behavior, which would serve to emphasize cryptocurrencies' behavior as analogous to established asset classes. Simple returns for discrete trading weeks and months are calculated together with the corresponding Pearson correlations for these time frames. Thereafter, correlation between returns and their respective correlation pairs are calculated for all three correlation measures (Spearman, Kendall, Pearson) and the mean correlation for all three measures is extracted for comparison.

6.2.2 Liquidity

We test for the liquidity criterion via two metrics, both spanning the previous 3 months leading up to December 8th, 2017.[2] First, we compare absolute daily trading volume of Bitcoin (BTC) and Ether (ETH) to the five stocks comprising the FAANG group, Facebook, Amazon, Apple, Netflix, Google/Alphabet. Thereafter, we compare the equally weighted averages of absolute daily liquidity for the top 10 cryptographic coins and top 10 cryptographic tokens by market cap to the FAANG basket's equally weighted daily liquidity. Second, we compare the ratio of daily trading volume to daily market capitalization for both the individual titles as well as the equally weighted baskets. Measures of comparison are the minimum, mean, maximum and standard deviation of values due to their significance to investment management, as well as the ratio between standard deviations and mean values for purposes of comparison.

6.2.3 Stability

Market Stability is tested as the resistance of cryptocurrency markets to trigger Market Wide Circuit Breaks (MWCB) and Limit Up-Limit Down levels (LULD) as an indicator for market maturity. For Market Wide Circuit Break rules, we use trigger levels and computations as established by the SEC filing Release No. 34–67,090 of the 31st of May 2012 as submitted by the Self-Regulatory Organizations (SRO)[3] (SEC 2012). MWCBs are specified as intraday market drops of an index of more than 7, 13 and 20% relative to the previous day's closing price, which are denoted as levels 1, 2 and 3 respectively. MWCBs lead the market to

[2]While alternative timeframes of 6, 9 and 12 months have been examined, results of the 3 month window are robust to such changes. Therefore, we limit our analysis to the three most recent months.

[3]The organizations participating in the SRO are: BATS Exchange, Inc.; BATS Y-Exchange, Inc.; NASDAQ OMX BX, Inc.; Chicago Board Options Exchange, Incorporated; C2 Options Exchange, Incorporated; Chicago Stock Exchange, Inc.; EDGA Exchange, Inc.; EDGX Exchange, Inc.; Financial Industry Regulatory Authority, Inc.; International Securities Exchange LLC; The NASDAQ Stock Market LLC; New York Stock Exchange LLC; NYSE Amex LLC; NYSE Arca, Inc.; National Stock Exchange, Inc. and NASDAQ OMX PHLX LLC.

halt for 15 minutes for levels 1 and 2 and to halt for the remainder of the trading day after breaking level 3. We construct a cryptocurrency index using some of the largest cryptocurrencies by market capitalization for which tick-by-tick data are available, representing 87% of total cryptocurrency market capitalization (for details, see separate annex).

Within our index, cryptocurrencies are weighted with their market capitalization relative to the total capitalization of cryptocurrencies included in the index. Individual weightings are rebalanced daily. Additionally, we run an MWCB test for Bitcoin alone due to the fact that Bitcoin represents 64% of total market capitalization of all existing cryptocurrencies as of the writing of this paper. We apply the MWCB criteria to our index and Bitcoin tick-by-tick time series to test for market breaks.

Limit Up-Limit Down levels are triggered by price moves of individual securities exceeding ±5%, 10% and 20% within a 5 min interval, if after triggering one of these levels the price of the individual security does not retract back below or above the threshold within 15 s. We apply the LULD criteria to Bitcoin due to its significance and position as an indicator for the larger crypto space. LULD trigger frequencies are approximated via fixed 5 min intervals yielding 288 discrete time windows per day. Tick-by-tick returns are calculated for each transaction against the first tick of the respective 5 min interval. After the passage of a 5 min interval, the incremental transaction defines the closing price against which to calculate returns for the following 5 min.

6.2.4 Portfolio Optimizations

Portfolio optimization proceeds in two steps. In the first step, we select four different portfolios composed exclusively of traditional asset classes, including stocks, bonds, real estate, gold and oil. The simplest allocation of portfolio 1 with only stocks and bonds (P1(B)) is extended by adding real estate (P2(B)), real estate and gold (P3(B)) and finally real estate, gold and oil (P4(B)). These are our benchmark portfolios, hence the notation (B). For each of the benchmark portfolios, we create a second version including the new asset class cryptocurrency represented by Bitcoin, which we label crypto portfolios (for details, see separate annex). We choose Bitcoin as a proxy for cryptocurrency due to its dominance of and correlation with the crypto sphere as a whole. Then, we optimize the Sharpe ratio of each benchmark and crypto portfolio retrospectively for each quarter using Excel's Solver.[4] Within the optimizations, negative asset class weightings (short positions) are permitted, except for constellations without convergence in solutions. Portfolio metrics include the daily returns:

$$r_{PF} = \mathrm{w} \cdot r$$

the portfolio variance:

[4]We also performed a check on Minimum-Variance optimization. Due to the high volatility levels of cryptocurrencies, we did not find any significant volatility reduction.

$$\sigma^2_{PF} = \mathrm{w} \cdot \Sigma \cdot w$$

and the Sharpe ratio for portfolios:

$$S_{PF} = \frac{r_{PF}}{\sigma_{PF}}$$

The risk-free rate for optimizations is fixed at 0%, firstly, because the risk-free rate for the time periods considered (Q2 2013 to Q3 2017) has been fluctuating around this marker and secondly, because we want to exclude interest rate effects from our estimations. This step in the analysis is not intended for purposes of investment advice but rather to evaluate how stable asset allocations will be over time and how large the contribution of cryptocurrency can be in the context of optimizations. This yields quarterly *ex-post* optimal asset class weightings for each portfolio.

The *ex-post* optimal weightings for each quarter of the first step are used in a second step to calibrate portfolio structures *ex-ante* for each following quarter. That is, the optimal weight for each preceding quarter is used to define portfolio allocation for the following quarter. With quarterly changing weights this approach is labelled *dynamic*. Due to wildly fluctuating portfolio weights for cryptocurrency under the dynamic approach we implement a second approach using the average of quarterly optimized weights for each asset class uniformly for all quarters, called the *averages* approach. Since after each quarter, asset allocations will have departed from the averages initially used due to positive or negative performance of the asset classes, allocations are rebalanced to their averages at the end of each quarter. For both the dynamic and averages approaches, allocation to cryptocurrency is comparatively high at 10% (for details, see separate annex). Therefore we implement a third approach where the proportion of cryptocurrency added to traditional portfolios is kept at a flat 1% for all four portfolios (for details, see separate annex). Hereby, the 1% share allocated to cryptocurrency is taken out of the allocation to equities due to the fact that the risk/return profile of cryptocurrency matches that of equities most closely compared to the other asset classes considered. Due to an emphasis on risk reduction, this approach is called *conservative*.

7 Results

7.1 Correlation

All cryptocurrency return time series are non-normal for both Lilliefors' and Stephens' modification of Kolmogorov-Smirnov, as well as Shapiro Wilk's *W*. Therefore, Spearman's rho seems most appropriate to evaluate correlation, showing strong correlation within the class of cryptocurrencies. 95% of cryptocurrency pairs

Correlation within Cryptocurrencies

	Spearman ρ		Kendall τ		Pearson r	
	Significant	Not Significant	Significant	Not Significant	Significant	Not Significant
negative	0	0	0	0	0	5
positive	180	10	183	7	140	45

Fig. 3 Clear statistically significant correlation within 190 cryptocurrency pairs, both for parametric and non-parametric tests

Instances of Correlation between a Cryptocurrency and a Traditional Asset Class/ Asset

Trad. Asset Class	Spearman ρ	Kendall τ	Pearson r	Consensus	of Total
Indices	**5**	5	9	**1**	160
Equities	**5**	5	11	**3**	180
Currencies	**1**	1	2	**0**	80
Bonds	**0**	0	3	**0**	20
Commodities	**6**	5	3	**0**	60
Real Estate	**1**	1	3	**1**	20
Total	**18**	17	31	**5**	**520**

Fig. 4 Only 1% of pairs (5 out of 520) between cryptocurrencies and traditional asset classes show statistically significant correlation over all three measures

showing significant positive correlation and therefore mutual necessity confirms cryptocurrencies as a coherent whole (Fig. 3).

Analyses of correlation between asset classes show that cryptocurrencies as a whole move independently of all traditional asset classes. For Spearman's rho, only 18 out of 520 pairs total show statistically significant correlation. When considering pairs with significant correlation over all three correlation measures, only five positively correlated pairs remain. Among those five, Bitcoin, as the dominant cryptocurrency to this date, shows only one instance of weak positive correlation to real estate with rho and r of 0.06 (Figs. 4 and 5).

The conclusions drawn are robust to adjustments in timeframes considered (for details, see separate annex). These results suggest a clear distinction of cryptocurrency as separate from traditional asset classes by fulfilling both the criterion of mutual exclusivity as well as differing returns.

Support for the classification of cryptocurrency specifically as an asset class is found in cryptocurrency's analogous behavior to traditional assets of increasing correlation for decreasing returns. This tendency shows striking similarity to behavior of the five FAANG stocks, both for discrete trading weeks as well as months (Fig. 6).

Moderate correlation between coins and tokens both measured via equally weighted daily mean and median returns do support the idea of two distinct sub asset classes. Results show robustness over time as evidenced by comparison of multiple 100 day slices (for details, see separate annex) (Fig. 7).

Following Sharpe's definition of an asset class in the evaluation of cryptocurrency, the three conditions of mutual exclusivity, exhaustiveness and

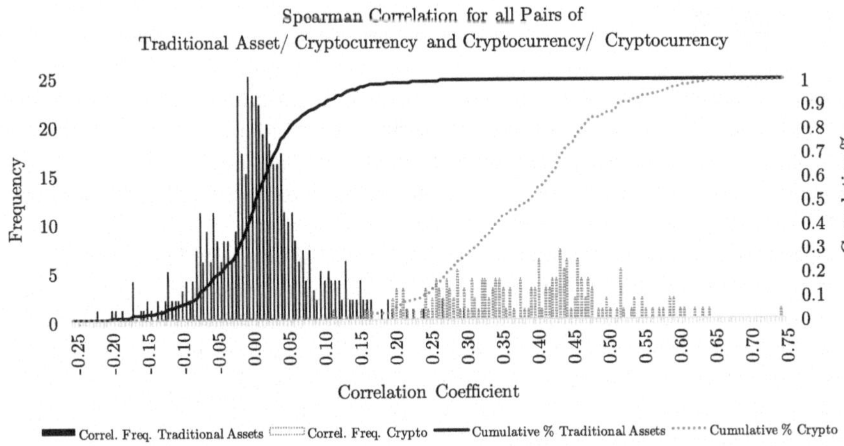

Fig. 5 Chart displays frequency and cumulative probability distribution of Spearman correlation for pairs of cryptocurrencies and traditional assets (blue) as well as pairs of cryptocurrencies (yellow). Cryptocurrencies and traditional assets largely uncorrelated, cryptocurrencies positively correlated among each other

Correlation of Asset Returns with Correlations of Asset Pairs

	Asset	Mean Spearman ρ	Mean Kendall τ	Mean Pearson r
By Trading Week	FAANG	-0.197	-0.132	-0.203
	Cryptocurrencies	-0.207	-0.140	-0.187
	Δ	0.010	0.008	0.016
	Asset	**Mean Spearman ρ**	**Mean Kendall τ**	**Mean Pearson r**
By Trading Month	FAANG	-0.453	-0.345	-0.412
	Cryptocurrencies	-0.414	-0.281	-0.205
	Δ	0.039	0.064	0.207

Fig. 6 FAANG stocks and cryptocurrencies both show increasing correlation for decreasing returns and vice versa. This supports cryptocurrencies as an asset class

Correlation between Coins and Tokens

Measure	Spearman ρ	Kendall τ	Pearson r
Mean Daily Returns	**0.36**	0.25	0.37
Median Daily Returns	**0.42**	0.29	0.44

Fig. 7 Both mean and median equal weight daily returns indicate only moderate correlation between coins and tokens

differing returns seem to be satisfied by our results. The findings of our correlation analyses therefore confirm cryptocurrency as a new asset class, while the liquidity and stability criteria will shed light onto the maturity of the crypto space.

7.2 Liquidity

In terms of absolute daily liquidity Bitcoin is squarely positioned in the midst of the group of FAANG. While the Ether is not quite on equal footing yet, it is closing in with just over one standard deviation difference to Netflix. Taking the equally weighted means of the FAANG basket, coin and token baskets highlights a substantial difference in maturity between the three groups. Here, FAANG clearly outperforms coins which in turn outperform tokens in absolute daily liquidity (Fig. 8).

Considering the ratio between daily liquidity and market capitalization, cryptocurrencies show significantly stronger liquidity compared to FAANG stocks across the board. Volatility of daily liquidity is highest for coins followed by tokens and FAANG stocks. Considering individual titles, both Bitcoin and Ether show stronger relative liquidity than all the components of the FAANG group. The most liquid traditional tech title, Netflix, is on average approximately half as liquid as Bitcoin relative to market capitalization. When considering the baskets of stocks and cryptocurrency, minimum relative liquidity of tokens (1.04%) is still higher than mean liquidity of the FAANG basket (0.73%), and minimum relative liquidity in coins (1.75%) outperforms even maximum daily liquidity of the FAANG basket (1.58%) (Fig. 9).

While in absolute terms both the Bitcoin and Ether can already compete with staple names in the equity sphere, the coins as a group still can't reach comparable trading volume. This emphasizes the relative dominance of the top two cryptocurrencies at this moment and points to relatively weak liquidity in smaller coins. However, the fact that beyond Bitcoin there are in fact further coins within only one standard deviation to the group of FAANG stocks shows that cryptocurrencies are in the process of catching up to legacy tech titles. Cryptographic

Daily Trading Volumes Sep. to Dec. 2017:
Equally Weighted Index & Individual

	Mean US$	Max US$	St.Dev. US$	St.Dev.\Mean
Top 10 Tokens	9,405,893	32,388,998	6,805,623	72.4 %
Top 10 Coins	582,942,401	2,089,626,410	422,109,317	72.4 %
FAANG	2,819,931,843	7,503,183,600	993,767,795	35.2 %
BTC	3,105,646,815	12,656,300,000	2,248,090,467	72.4 %
ETH	764,654,761	2,675,940,000	485,548,077	63.5 %
Facebook	2,777,700,092	7,459,686,000	1,330,500,617	47.9 %
Amazon	3,769,808,831	17,942,740,000	2,356,693,359	62.5 %
Apple	4,688,643,769	11,599,890,000	1,961,176,666	41.8 %
Netflix	1,323,544,711	4,787,650,000	812,804,793	61.4 %
Alphabet\Google	1,539,961,812	5,412,848,000	659,235,131	42.8 %

Fig. 8 Bitcoin on equal footing with FAANG equities, Ether closing in

Daily Trading Volume to Market Cap Ratio
Sep. to Dec. 2017: Equally Weighted Index & Individual

	Min %	Mean %	Max %	St.Dev. %	St.Dev.\Mean
Top 10 Tokens	1.04	2.58	5.73	1.25	48.3 %
Top 10 Coins	1.75	5.21	18.20	3.01	57.8 %
FAANG	0.39	0.73	1.58	0.25	34.1 %
BTC	1.49	2.99	7.91	1.31	43.8 %
ETH	0.91	2.51	9.51	1.46	58.3 %
Facebook	0.28	0.55	1.44	0.26	47.6 %
Amazon	0.42	0.74	3.38	0.42	57.1 %
Apple	0.27	0.56	1.40	0.23	41.3 %
Netflix	0.50	1.58	5.56	0.94	59.7 %
Alphabet\Google	0.11	0.22	0.76	0.09	41.0 %

Fig. 9 Significantly stronger liquidity and volatility of liquidity in cryptocurrencies

tokens still seem to be in their infancy stages and will need to grow in order to prove a serious investment alternative, at least in terms of liquidity.

While the spheres of coins and especially tokens on average still show a need for volume growth, in absolute terms liquidity relative to market capitalization is very strong, outperforming the staple equities by orders of magnitude. How this ratio will evolve with increasing maturity of the sector remains to be seen.

Volatility of individual cryptocurrency liquidity is in line with FAANG equities despite being at the top end for both absolute and relative measures. Comparing the basket of FAANG equities to baskets of coins and tokens, volatility of liquidity is decidedly higher in cryptocurrencies. This is illustrated by column 6 of Table 8 above, which shows the ratio between the standard deviation and mean for the ratio between daily liquidity and market capitalization. The pronounced difference between cryptocurrencies and FAANG stocks might mainly be traced back to mitigating averaging effects among FAANGs which seem to be absent among cryptocurrencies. That is, particularly high or low relative daily liquidity in individual FAANGs seems to be compensated by the basket as a whole, while no compensating effect seems to occur within the basket of cryptocurrencies, indicating correlation among individual cryptocurrency liquidity.

Our results point to the conclusion that despite its infancy, the cryptocurrency sphere already possesses significant liquidity. Both robust absolute daily liquidity as well as strong ratios between trading volume and market capitalization point to the conclusion that the cryptocurrency sphere does provide the necessary liquidity for investment positions to be entered and adjusted effectively, especially when compared to FAANG stocks as the benchmark.

7.3 Stability

Bitcoin, representing about two thirds of total cryptocurrency market capitalization as of Dec. 2017, shows rather unstable behavior, with a total of 92 Market Wide Circuit Breaks over the years 2016 and 2017. In comparison, MWCB have only been triggered once in the US since their inception in 1988 (Ackert 2012). Interestingly, MWCB events increase in frequency from 2016 to 2017, possibly due to intermediate overdue price corrections. Market breaks are less frequent for our cryptocurrency index (Fig. 10).

Two level 3 breaks following an intraday drop of −20% or more are eliminated through diversification. However, this cannot alleviate the fact that a comparatively high amount of market breaks occur in cryptocurrencies and especially in Bitcoin, which repeatedly points to a need for maturation of the sector (Fig. 11).

Limit-Up Limit-Down triggering, as an indicator for wild short-term fluctuations within a 5 min interval, has so far been comparably prevalent in Bitcoin. However, the quantity of trigger signals has steadily decreased since Q1 2016, indicating a gradual increase in stability. Generally, short term volatility is higher in the first half of a year compared to the second half.

The interplay of a decreasing quantity of Limit-Up Limit-Down trigger signals concurrent with an increase in Market Wide Circuit Breaks points to decreased short-term volatility clustering possibly at the expense of higher longer-term volatility. While these results can be interpreted as a step toward the smoothing out of market behavior and therefore maturation of the sector, investors in the asset class cryptocurrency will, at least for the time being, encounter significant 'bumps in the road'.

BTC: Number of Days with MWCB

		Level 3 -20%	Level 2 -13%	Level 1 -7%
2016	Q1		2	6
	Q2		2	5
	Q3	1	1	3
	Q4			1
2017	Q1	1	4	14
	Q2	2	4	16
	Q3	1	1	19
	Q4		2	7
Total		**5**	**16**	**71**

INDEX: Number of Days with MWCB

		Level 3 -20%	Level 2 -13%	Level 1 -7%
2016	Q1		2	4
	Q2		1	6
	Q3	1	1	3
	Q4			1
2017	Q1	1	3	11
	Q2	1	4	13
	Q3		1	15
	Q4		1	6
Total		**3**	**13**	**59**

Fig. 10 Both BTC and our cryptocurrency index show significant volatility and quantity of market breaks. Index slightly more stable

BTC: Number of Limit-Up Limit-Down Triggers

		±20%		±10%		±5%		Total		
		Pos	Neg	Pos	Neg	Pos	Neg	20%	10%	5%
2016	Q1			2	5	12	10		7	22
	Q2					3	1			4
	Q3					1	1			2
	Q4									
2017	Q1					7				7
	Q2					6	2			8
	Q3									
	Q4					1	3			4
Total		**0**	**0**	**2**	**5**	**30**	**17**	**0**	**7**	**47**

Fig. 11 Limit-Up-Limit-Down triggering reinforces picture of instability

7.4 Portfolio Optimization

For nearly all quarters between Q2 2013 and Q3 2017, we find a positive impact from adding cryptocurrency to portfolios when considering the *ex-post* Sharpe ratio (for details, see separate annex). Adding cryptocurrency does not improve minimum variance portfolio structures, as expected. While these results are a good *ex-post* indicator for how a portfolio should have been allocated in hindsight it does not follow that these allocations automatically deliver outperformance if used as a rule to calibrate portfolios *ex-ante* for the following quarter (Fig. 12).

For the implementation of our *ex-post* weights in *ex-ante* calibration, we find distinct results for each of our three approaches (dynamic, averages, conservative). The dynamic approach, using each past quarter's optimized weights for the allocation in the following quarter, leads to underperformance of crypto portfolios compared to traditional portfolios. Portfolio variance is higher for portfolios one, two, and three when compared to purely traditional portfolios. Moreover, all crypto portfolios underperform traditional portfolios when considering the Sharpe ratio. Not only do portfolios structured this way underperform but allocation to cryptocurrency is (a) comparatively high at an average of 10% and (b) fluctuating wildly with an average range of 107% across portfolios. While an average portfolio weight of 10% might be tolerable, the significant variability raises doubts as to the implementability of such portfolios. Thus, in a next step, we test our second approach using average cryptocurrency allocation of about 10% while eliminating the variability of the allocation.

Implementing the averages approach, using the average of quarterly weights for each asset class with rebalancing to initial weights at the end of each quarter, delivers the best results with significant and persistent outperformance of crypto portfolios

Quarterly Ex-Post Allocation to Cryptocurrency

	Mean	Min	Max	Range
P1: Stocks, Bonds	9.96%	-11.45%	95.35%	106.80%
P2: Stocks, Bonds, Real Estate	10.02%	-10.95%	95.35%	106.30%
P3: Stocks, Bonds, Real Estate, Gold	9.94%	-10.61%	95.35%	105.96%
P4: Stocks, Bonds, Real Estate, Gold, Oil	9.63%	-13.31%	95.35%	108.66%
Mean	**9.89%**	**-11.58%**	**95.35%**	**106.93%**

Fig. 12 High 10% mean and 107% range of allocation to cryptocurrency across portfolios

Effects of adding Cryptocurrencies to Traditional Portfolios
Ex-Ante Simulation Results

		Dynamic	Averages	Conservative
P1: Stocks, Bonds	Δ Daily σ	0.63%	0.81%	0.03%
	Δ Sharpe	-45.29%	37.12%	26.24%
P2: Stocks, Bonds, Real Estate	Δ Daily σ	0.61%	0.80%	0.03%
	Δ Sharpe	-44.31%	43.43%	25.97%
P3: Stocks, Bonds, Real Estate, Gold	Δ Daily σ	0.59%	0.78%	0.03%
	Δ Sharpe	-53.33%	43.56%	29.65%
P4: Stocks, Bonds, Real Estate, Gold, Oil	Δ Daily σ	-0.18%	0.57%	0.03%
	Δ Sharpe	-20.87%	50.08%	50.62%

Fig. 13 Averages and Conservative portfolios strongly outperform benchmark for risk-adjusted returns

over benchmark portfolios (Fig. 13). Crypto portfolios remain more volatile than traditional portfolios. However, the averages mechanism consistently outperforms traditional portfolios in terms of the Sharpe ratio, improving the ratio on average by 43.5%. Adding additional asset classes into portfolios going from P1 to P4 increases outperformance of crypto portfolios (Fig. 14).

Due to the fact that an allocation of 10% to a newly emerging asset class remains comparatively high in light of typical investment management practice, we finally

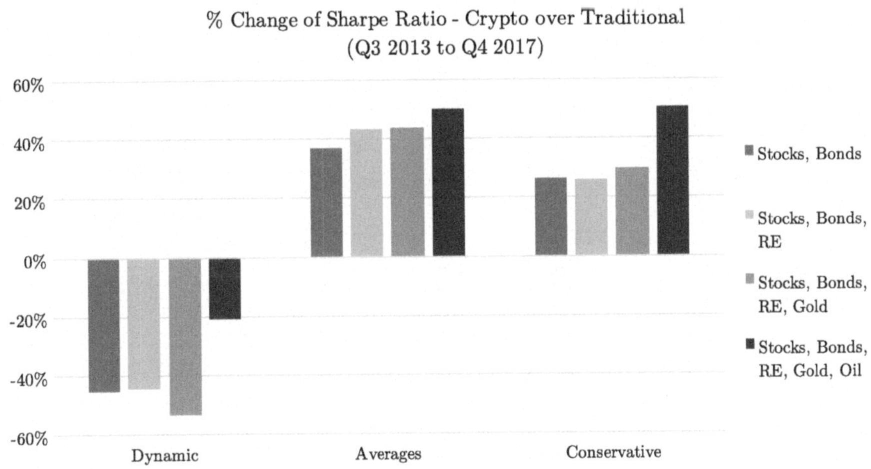

Fig. 14 Averages approach with strongest risk-adjusted outperformance

implement our conservative approach with a flat 1% allocation to cryptocurrency. We find that the conservative crypto portfolios can maintain outperformance in terms of the Sharpe ratio, albeit at a lower level than averages portfolios. Moreover, the gap between crypto portfolios and traditional portfolios in terms of volatility narrows significantly to only 0.03%. Risk-adjusted outperformance again generally increases when adding asset classes from P1 to P4.

Our results show that using Sharpe optimized *ex-post* weights for quarter-by-quarter *ex-ante* cryptocurrency asset allocation leads neither to reduced portfolio volatility nor to risk-adjusted outperformance. However, keeping the allocation to the asset class cryptocurrency stable at the average of optimized quarters causes volatility to increase, but also leads to strong risk-adjusted outperformance of crypto portfolios that is the best of all approaches implemented. The conservative approach also produces outperformance, albeit not as strongly as the averages approach. However, portfolio volatility is nearly similar to that of purely traditional portfolios.

Equity curves for the best performing averages approach highlight that while P1 to P3 deliver higher absolute returns than P4, the combination of cryptocurrency with stocks, bonds, real estate, gold and oil in P4 shows strongest risk-adjusted performance of all portfolios (Fig. 15).

These findings suggest that conservatively supplementing traditional portfolio structures with cryptocurrency, while not a free lunch, appears to be a risk-effective way of increasing portfolio performance. For investors with an increased risk budget, using averages of ex-post Sharpe optimized weights might provide an effective way to maximize risk-adjusted returns. However, our results also suggest that with increasing allocation to cryptocurrency there might be a tipping point of negative marginal utility beyond which crypto portfolios will generate inferior results. With cryptocurrency, as is often the case, the poison might ultimately be in the dose.

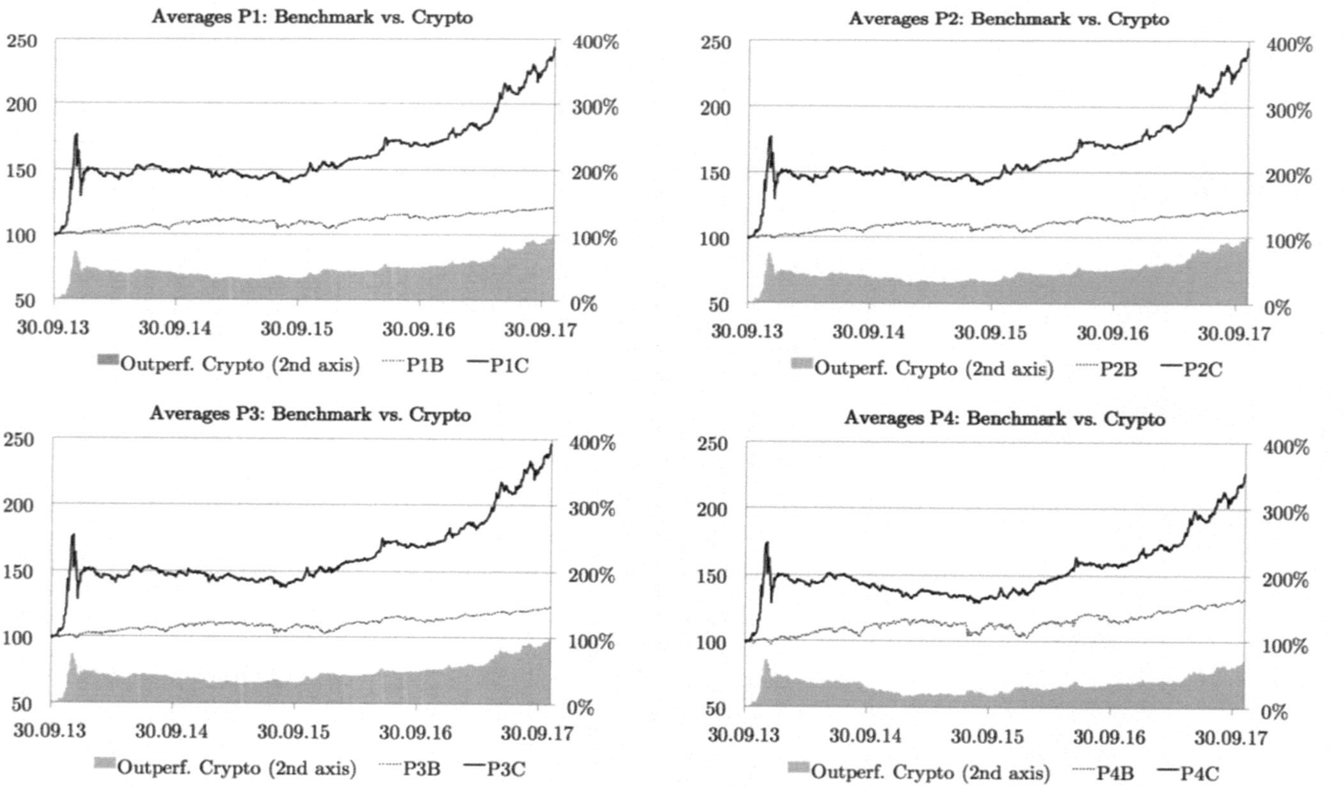

Fig. 15 P1 to P3 with very similar return curves. P4 weakest in absolute returns but with strongest risk-adjusted outperformance

8 Conclusion

We find that cryptocurrencies constitute a new distinct asset class and that supplementing traditional portfolios with cryptocurrency can lead to significant and persistent outperformance in risk-adjusted returns within the scope of our analyses.

Cryptocurrency qualifies as a distinct asset class by exhibiting high correlation among individual cryptocurrencies while being mostly uncorrelated with all traditional asset classes, aligning along Sharpe's (1992) asset class criteria of mutual exclusivity, exhaustiveness and differing returns. Absolute liquidity of Bitcoin is already on equal footing with FAANG equities, the Ether is closing in, while the remaining crypto space can not yet match the liquidity of traditional assets. Relative to their market capitalization, cryptocurrencies show significantly higher liquidity than the FAANG equities, both for coins and tokens. However, at least in terms of market stability there is still significant need for maturation of cryptocurrencies. With frequent Market Wide Circuit Break signals and, although decreasing, still a significant amount of Limit-Up Limit-Down interruptions, cryptocurrency trading would today remain rather discontinuous, were the rules of traditional equities exchanges applied.

Quarterly optimization of four traditional portfolio structures with and without cryptocurrency shows that adding cryptocurrencies to portfolios reliably improves quarterly *ex-post* Sharpe ratios while failing to reduce portfolio volatility. Turning to the analysis whether such *ex-post* weights lead to superior performance when implemented via *ex-ante* asset allocation in portfolios, we use three approaches. First, *ex-post* optimized portfolio weights of previous quarters are used for asset allocation in each following quarter (our dynamic approach) but fail to improve volatility or the Sharpe ratio. Second, asset allocation for each asset class using the average quarterly weight over all previously optimized quarters with quarterly rebalancing (the averages approach) leads to the strongest outperformance regarding Sharpe ratios but increases portfolio volatilities. Asset allocation to cryptocurrencies under the dynamic approach swings wildly between −10.61% and 95.35% with a mean of 10%, which is uniformly used as the basis for the averages approach. Finally, reducing the allocation to cryptocurrency to a conservative 1% for all quarters (our conservative approach), in order to accommodate risk consciousness of investment management practice, we find nearly similar volatilities of crypto portfolios and traditional portfolios paired with strong outperformance in Sharpe ratios which, however, can't match the outperformance of our averages approach.

Our results suggest that there is significant upside to be captured by investment practitioners from the careful addition of cryptocurrency to traditional portfolio structures. Comparatively conservative addition to otherwise conventional portfolio structures leads to persistent risk-adjusted outperformance. However, this requires a certain absolute level of risk appetite. Cryptocurrency consistently exhibits both significant short-term volatility clustering and an increase in long-term portfolio volatility, as evidenced by our stability analyses. In the future, possibilities for intra-asset class diversification regarding cryptocurrencies might develop as one means to improve the inherent risk profile. However, our results suggest that such effects are so far rather negligible. This might change in the future when individual cryptocurrencies might be differentiated based on specific value propositions tied to use cases or backing media of individual cryptocurrencies. Such differentiation might lead to a decoupling of return synchronicity and therefore greater potential to mitigate individual coin volatility. Investment professionals may want to look out for changing correlation and return dynamics as the asset class continues on its path to maturation.

Acknowledgements We thank Christian Gombert of the Hauck & Aufhäuser Privatbankiers as well as Johannes Bernius for their invaluable support with data acquisition. Lars Geiger, the participants of the 6th Crowdinvesting Symposium, and the participants of the 26th Annual Conference of the Multinational Finance Society have provided valuable feedback.

References

Ackert LF (2012) The impact of circuit breakers on market outcomes. Technical report. Economic Impact Assessment EIA9. UK Government Office for Science – Foresight Project. https://www.gov.uk/government/uploads/system/uploads/attachment_data/file/289039/12-1070-eia9-impact-circuit-breakerson-market-outcomes.pdf

Allen M (2017) Chiasso accepts tax payments in Bitcoin. swissinfo.ch. https://www.swissinfo.ch/eng/business/swiss-fintech_chiasso-accepts-taxpayments-in-bitcoin/43503464

Back A (2002) Hashcash-a denial of service counter-measure. ftp://sunsite.icm.edu.pl/site/replay.old/programs/hashcash/hashcash.pdf

Baird L (2016) The Swirlds hashgraph consensus algorithm: fair, fast, Byzantine fault tolerance. Technical report Swirlds-Tr-2016-01

Briere M, Oosterlinck K, Szafarz A (2015) Virtual currency, tangible return: portfolio diversification with Bitcoin. J Asset Manag 16(6):365–373

Buterin V (2013a) A next-generation smart contract and decentralized application platform. Technical report. Ethereum project white paper. https://github.com/ethereum/wiki/blob/master/drafts/%5Benglish%5D-old-ethereumwhitepaper.md

Buterin V (2013b) Bitcoin network shaken by blockchain fork. bitcoinmagazine.com. https://bitcoinmagazine.com/articles/bitcoin-network-shaken-by-blockchainfork-1363144448/

Buterin V (2016) "Ethereum 2.0". Ethereum project mauve paper. https://cdn.hackaday.io/files/10879465447136/Mauve%20Paper%20Vitalik.pdf

Buterin V (2017a) On medium-of-exchange token valuations. http://vitalik.ca/general/2017/10/17/moe.html

Buterin V (2017b) The meaning of decentralization. Technical report. Medium. https://medium.com/@VitalikButerin/the-meaning-of-decentralization-a0c92b76a274

CBOE (2017) XBT-Cboe Bitcoin futures. Technical report. Chicago Board Options Exchange. http://cfe.cboe.com/cfe-products/xbt-cboe-bitcoin-futures

CME (2017) Now available: Bitcoin futures. cmegroup.com. https://www.cmegroup.com/trading/bitcoin-futures.html

Dougherty C, Huang G (2014) Mt. Gox seeks bankruptcy after 480 million Bitcoin loss. bloomberg.com. https://www.bloomberg.com/news/articles/2014-02-28/mt-gox-exchange-files-for-bankruptcy

Dwork C, Naor M (1992) Pricing via processing or combatting junk mail. In: Annual international cryptology conference. Springer, Heidelberg, pp 139–147. https://link.springer.com/chapter/10.1007/3-540-48071-4_10

Dyhrberg AH, Foley S, Svec J (2018) How investible is Bitcoin? Analyzing the liquidity and transaction costs of Bitcoin markets. Econ Lett 171:140–143

Eisl A, Gasser S, and Weinmayer K (2015) Caveat emptor: does Bitcoin improve portfolio diversification? Available at SSRN 2408997

Elendner H et al (2018) The cross-section of crypto-currencies as financial assets - investing in crypto-currencies beyond bitcoin. In: Handbook of blockchain, digital finance, and inclusion, vol 1. Elsevier, Amsterdam, pp 146–170

EOSProject (2017) EOS.IO technical white paper. Technical report EOS Project. https://github.com/EOSIO/Documentation/blob/master/TechnicalWhitePaper.md

Fink C, Johann T (2014) Bitcoin markets. Available at SSRN 2408396

GolemProject (2016) The Golem project - crowdfunding whitepaper. Technical report. The Golem Project. http://golemproject.net/doc/DraftGolemProjectWhitepaper.pdf

Higgins S (2014) Hal Finney on Bitcoin: in his own words. coindesk.com. https://www.coindesk.com/hal-finney-bitcoin-words/

Jakobsson M, Juels A (1999) Proofs of work and bread pudding protocols. In: Secure information networks. Springer, Berlin, pp 258–272. https://link.springer.com/content/pdf/10.1007/978-0-387-35568-9_18.pdf

Johnston D et al (2015) The general theory of decentralized applications, DApps. GitHub 9. https://github.com/TarantulaTechnology/Documents-Blockchain/blob/master/The%20General%20Theory%20of%20Decentralized%20Applications%2C%20DApps.pdf

Juels A, Brainard JG (1999) Client puzzles: a cryptographic countermeasure against connection depletion attacks. In: NDSS, vol 99, pp 151–165. http://www.isoc.org/isoc/conferences/ndss/99/proceedings/papers/juels.pdf

Kalla S (2017) A framework for valuing crypto tokens. Technical report. Acupacy. https://www.coindesk.com/framework-valuing-crypto-tokens/

Lee DKC, Li G, Wang Y (2018) Cryptocurrency: a new investment opportunity? J Altern Invest 20 (3):16–40

Merkle RC (1980) Protocols for public key cryptosystems. In: Security and privacy, 1980 IEEE Symposium on. IEEE, pp 122–134. https://www.researchgate.net/profile/Ralph_Merkle/publication/220713913_Protocols_for_Public_Key_Cryptosystem/links/00b495384ecda07784000000/Protocols-for-Public-Key-Cryptosystems.pdf

Merkle RC (1990) A certified digital signature in Conference on the theory and application of cryptology. In: Brassard G (ed) Advances in cryptology. CRYPTO'89 LNCS 435. Springer, Heidelberg, pp 218–238. https://link.springer.com/content/pdf/10.1007/0-387-34805-0_21.pdf

Nakamoto S (2008) Bitcoin: a peer-to-peer electronic cash system. Technical report. https://bitcoin.org/bitcoin.pdf

P4Titan (2014) Slimcoin - a peer-to-peer crypto-currency with proof-of-burn. Technical report. slimcoin

Popov S (2017) The tangle. Technical report. IOTA Project. https://iota.org/IOTA_Whitepaper.pdf

ProtocolLabs (2017) Filecoin: a decentralized storage network. Technical report. Protocol Labs. https://filecoin.io/filecoin.pdf

Rosenfeld M (2012) Analysis of hashrate-based double spending. arXiv preprint arXiv:1402.2009. https://arxiv.org/pdf/1402.2009.pdf

SEC (2012) Approval order. Technical report Release 34-67090. Securities and Exchange Commission. https://www.sec.gov/rules/sro/bats/2012/34-67090.pdf

Sharpe WF (1992) Asset allocation: management style and performance measurement. J Portf Manag 18(2):7–19

Swissinfo (2016) Zug first to accept Bitcoin for government services. swissinfo.ch. https://www.swissinfo.ch/eng/business/crypto-valley_zug-firstto-accept-bitcoin-for-government-services/42143908

Szabo N (2008) Bit gold. http://unenumerated.blogspot.de/2005/12/bitgold.html

Trimborn S, Li M, Härdle WK (2018) Investing with cryptocurrencies - A liquidity constrained investment approach

Ummelas O (2018) Estonia scales down plan to create national cryptocurrency. bloomberg.com. https://www.bloomberg.com/news/articles/2018-06-01/estonia-curbs-cryptocurrency-plan-that-drew-rebuke-from-draghi

Voshmgir S, Kalinov V (2017) Blockchain - A beginners guide. Technical report Version 1.0. BlockchainHub. https://blockchainhub.net/blockchaintechnology

Wei WC (2018) Liquidity and market efficiency in cryptocurrencies. Econ Lett 168:21–24

WeiDai (1998) B-Money. http://www.weidai.com/bmoney.txt

Wood G (2014) Ethereum: a secure decentralised generalised transaction ledger. Ethereum project yellow paper EIP-150 revision. http://www.cryptopapers.net/papers/ethereum-yellowpaper.pdf

Are Virtual Currencies Virtuous? Ethical and Environmental Issues

Sondes Mbarek, Donia Trabelsi, and Michel Berne

Abstract Cryptocurrencies have gained in popularity and generated a great deal of enthusiasm in recent years with regard to the sustained increase in the number of transactions achieved by miners. On what scale can we consider the process and uses of virtual money to be ethical? What are the misuses related to their use? In this chapter, we study the ethical and environmental issues of cryptocurrencies. First, regarding the environmental issue, the major cryptocurrencies use a large amount of electricity for mining, which has a significant impact on the energy production system and global warming. Second, we discuss the new type of Dark economy that has emerged with these currencies, thanks to the anonymity of transactions. We particularly emphasize the unethical use of cryptocurrencies, namely the virtual money laundering and tax evasion, the financing of illegal activities (i.e. illicit products, terrorist financing) and cyber-attacks. Third, we develop the ethical use of virtual money and show that this kind of currency, which guarantees the protection of privacy and anonymity of transactions, can be a good solution to mitigate transaction costs and reduce poverty. They can also be beneficial in the context of debt crises and hyperinflation. Thus, cryptocurrencies per se are not evil; it is their uses that can be.

1 Introduction

When we talk about cryptocurrencies, of course we have in mind Bitcoin and the enthusiasm it has generated in recent years among investors, media and academics. Rightly so, between October 2016 and October 2017, Bitcoin's market capitalization increased from $10.1 to $79.7 billion and its price from $616 to $4800, representing an annual profitability of 680% (Corbet et al. 2019). This is an unusual, if not exceptional, profitability for an asset. Dyhrberg (2016) also shows that Bitcoin can be used as a hedge against the shares of the Financial Times Stock Exchange Index and against the US dollar in the short term. Bitcoin would even have some of the same hedging capabilities as gold and can be included in the variety of tools available to

S. Mbarek (✉) · D. Trabelsi · M. Berne
LITEM, Univ Evry, IMT-BS, Université Paris Saclay, Evry, France
e-mail: sondes.mbarek@imt-bs.eu; donia.trabelsi@imt-bs.eu; michel.berne@imt-bs.eu

S. Goutte et al. (eds.), *Cryptofinance and Mechanisms of Exchange*, Contributions to Management Science, https://doi.org/10.1007/978-3-030-30738-7_2

market analysts to hedge specific market risks and become "the" new safe haven security. More recently, Urquhart and Zhang (2018) assess the relationship between Bitcoin and currencies and conclude that Bitcoin can provide intraday hedging for CHF, EUR and GBP, but acts as a diversifier for AUD, CAD and JPY. They also argue that Bitcoin is a safe haven during periods of extreme market turbulence for CAD, CHF and GBP. Guesmi et al. (2018), on the other hand, analyze the cross-conditional effects and volatility spillovers between Bitcoin and other financial assets, demonstrating that bitcoin can offer undeniable diversification benefits and hedging opportunities for investors. In particular, the results show that hedging strategies involving gold, oil, emerging stock markets and Bitcoin significantly reduce the variance of a portfolio compared to the variance of a portfolio composed solely of gold, oil and equities. Virtual currencies also offer many potential benefits, including faster and more efficient payments and transfers—especially across borders—and ultimately the promotion of financial inclusion (IMF 2016). Moreover, cryptocurrencies can provide solutions to social problems. In fact, they can help reduce poverty, transaction costs and hyperinflation while ensuring the anonymity of transactions and protection of privacy (Vigna and Casey 2015; Dopfer et al. 2004; Maurer et al. 2013).

While the technological and financial implications of cryptocurrencies and more specifically Bitcoin have attracted much attention (Corbet et al. 2019), we have decided in this chapter to study other aspects of cryptocurrencies, namely ethical and environmental aspects. There is very little debate in the academic literature on these aspects (Dierksmeier and Seele 2018). For example, Corbet et al. (2019) conducted a review of the empirical literature on cryptocurrencies as a financial asset. Only 11 of the 92 studies reviewed, deal with cryptocurrencies and cybercrime, but none deals with other aspects related to the ethics and/or environmental impact of cryptocurrencies.

However, the question of the ethics of cryptocurrencies is closely linked to the very nature and functioning of these assets. Indeed, cryptocurrencies are not associated with any government authority or institution (Australian Transaction Reports and Analysis Centre (AUSTRAC) 2012, p. 8; Angel and McCabe 2015; Dierksmeier and Seele 2018). The Bitcoin system was explicitly designed to avoid relying on traditional trusted intermediaries, such as banks. The value of Bitcoin is not based on a tangible asset (traditional currencies, precious metals or other physical commodities) or on a country's economy but on the trust and honesty of its users (Nica et al. 2017) and on the security of an algorithm, a kind of public ledger called "Blockchain" which is able to track all transactions (Corbet et al. 2019). As such, they are allocated to miners as a reward for being the first to solve the mathematical problems necessary to add a new block of transactions to the blockchain (Angel and McCabe 2015).

According to the AUSTRAC report (2012, p. 9), due to their nature and the lack of strict regulations, cryptocurrencies are likely to attract criminal groups and individuals who seek to use them as an instrument to pay for illicit goods and services and hide the source of illicit funds or avoid taxation (Nica et al. 2017; AUSTRAC 2012). Corbet et al. (2019) define a trilemma specific to cryptocurrencies that exists between the potential for illicit use due to their anonymity, the lack of regulatory oversight and infrastructure gaps influenced by the growth of cybercrime. These digital currencies also pose challenges for government agencies to follow the money trail. Several services offering an enhanced anonymization of transactions have emerged in the Bitcoin ecosystem. Some of these services routinely process the equivalent of a six-digit dollar amount. In a series of experiments, Möser et al. (2013) used reverse engineering methods to understand how it works and to try to find anonymous transactions. Their results show that it is unlikely that a Know-Your-Customer (KYC) principle can be applied in the Bitcoin system. Indeed, Nica et al. (2017) argue that this association with crime has long led to considerable skepticism about cryptocurrencies among financial authorities, governments and the public. The "dark" side of cryptocurrencies has been widely covered in the media (Beigel 2018; Krugman 2013). In an article published on the New York Times blog, Nobel Prize winner Paul Krugman (2013) said that "Bitcoin is an evil", citing the argument that Bitcoin is part of a political agenda aimed at undermining central banks and the ability of governments to collect taxes.

While risks to the conduct of monetary policy appear less likely to occur at this stage given the very small size of virtual currencies, risks to financial stability may emerge as new technologies become more widely used (IMF 2016). This is also the conclusion of Aldridge et al. (2014) and Plassaras (2013) who argue that cryptocurrencies do not pose a threat to the financial or macroeconomic stability and would only pose risks if they were used substantially in several sectors of the economy. However, due to their digital nature and the ease of their global distribution, cryptocurrencies may be more ubiquitous than any other form of currency previously established (Dierksmeier and Seele 2018).

Besides the ethical dimensions listed above, the major cryptocurrencies have a significant environmental impact due to the large amount of electricity needed for mining. Energy production and use have ethical dimensions in terms of access to resources, pollution generation and global warming. Sovacool et al. (2013) use the concept of "energy justice", the fair dissemination both of the benefits and costs of energy services. Externalities, climate change, rising prices, corruption and social conflicts, uneven development and a burden on the poor can all be associated with energy injustice, which is rising with very large energy production and use.

Therefore, it would be useful to examine in more detail the ethical and environmental risks inherent in their use so that different users and stakeholders can make informed decisions. In addition, while the future of e-commerce involves a transition to digital currencies, it is essential that economic, political and legal institutions should be prepared (Plassaras 2013).

In this chapter, we review the risks associated with the use of cryptocurrencies, particularly from an ethical and environmental point of view. We explore the environmental implications of cryptocurrencies (Sect. 1) as well as their obscure aspects (Sect. 2). While it is true that a system in itself may not be bad, it can nevertheless be used unethically. For this reason, Sect. 3 is dedicated to the study of the "moral goods" of cryptocurrencies.

2 Cryptocurrencies and Energy Consumption: Virtual Currency but Real Environmental Impact

This section is devoted to a non-technical presentation of the links between cryptocurrency mining and energy consumption. Its aim is to discuss the impact of mining on the world energy production system and on global warming, as it exists in 2019.

The cryptocurrency world is dominated by Bitcoin followed by Ethereum. These two major currencies share a large number of features, the most notable of them being that they use the proof of work system of block validation. That is to say, when transactions happen, they are packaged into blocks and each block is validated by miners. The miners produce a block signature that shows that the block is correct and then the valid block is added to the existing blockchain. The proof of work procedure requires complicated computations, using a trial and error procedure and is therefore extremely energy-intensive. Independent miners compete to be the first to validate the block as this brings them a reward. Other validating systems exist (such as the proof of stake that we will discuss later) but the proof of work has proved to be absolutely secure up to now. Indeed, anybody wanting to tamper with a block content (for example, to change a past transaction) would have to conduct these intensive computations again and, as there are many copies of the blockchain, it is virtually impossible to reverse the existing consensus on the validity of a block.

Mining is not the only source of energy consumption in the Bitcoin and Ethereum systems, but the other ones are minor in comparison: as transactions and blocks are added to the blockchain, it becomes longer. According to Blockchain Luxemburg S.A. (2019), the size of the Bitcoin blockchain on May 5th, 2019 was 217 MB and growing at a steep rate of nearly 40% per year. Given the fact that many copies of the blockchain exist, this is definitely using up electricity for storage. Another use of energy is related to the rest of the cryptocurrencies' ecosystems, such as wallets.

2.1 *Why This Is a Serious Issue?*

To answer this question, it would be wise to make first an overview of the situation regarding cryptocurrencies in general (2.1.1) and then to focus on Bitcoin (2.1.2).

2.1.1 Bitcoin and Other Main Crypto Currencies: An Overview

With the development of Bitcoin and Ethereum, mining has become enormously energy-intensive. But actual energy consumption figures are not available and methodologies have been designed to provide a reasonable estimate. There are two main methodologies. Their basic idea is quite simple but choosing the right figures to feed the model is tricky and different assumptions will lead to different outcomes.

The first methodology, used by O'Dwyer and Malone (2014), starts with the estimation of the difficulty of block validation, which is known to a good degree of accuracy. Then it makes assumptions about the computing equipment used by miners and finds the energy consumption associated with the mining of one block.

The Digiconomist method (2019), as described on its website, following Hayes (2015a, b) assumes miners spend most of their income on energy. Their income is known as we will see below. Knowing the price of electricity, energy consumption is derived.

Whatever the methodology used, the figures show that the total electricity consumption due to Bitcoin mining is extremely high. For bitcoin, according to the Digiconomist (2019), the peak was between 60 and 73 TWh in October 2018, on an annualized basis. On May 6th, 2019, it was between 39 and 59 TWh, that is 0.27% of the world's electricity consumption or about the energy consumption of Colombia.

These figures are even more spectacular when given per transaction: on May 6th, 2019, the figure was 432 KWh. A popular comparison is often made with the Visa electronic payment system, which needs an energy consumption of more or less 300,000 times lower than Bitcoin for each transaction. The Bitcoin and Visa figures cannot be strictly compared but the general conclusion holds that Bitcoin transactions are extremely energy intensive compared to common electronic transactions. This has led Mora et al. (2018) to say that, if the existing trend lasts, cryptocurrency mining (mainly Bitcoin), will someday push global warming beyond 2 °C. While this is unlikely, as we will see later, this shows the degree of alarm linked to this topic. Others like Dilek and Furuncu (2019) add that this electricity consumption, mainly from fossil fuel, has a major impact on air pollution and therefore on human health. Finally, the process adds to the pile of e-waste (González 2016).

Figures for Ethereum provided by Digiconomist (2019) show lower energy consumption, both in total and per transaction (early May 2019: 7 TWH on an annual basis, 29 KWh per transaction, around 0.03% of the world's electricity consumption). However, these figures are still much higher than for other electronic transactions systems.

These views are not held to be true by all Bitcoin observers. Some of them, like Bevand (2017), believe the Digiconomist (2019) estimates are way too high. Bevand's own estimates are lower, but remain in the same order of magnitude. Bevand and some others also deny Bitcoin mining is wasteful, claiming that the benefits derived from Bitcoin usage are high: for example, Wimbush (2018) argues that Bitcoin and Ethereum creation, through mining, has allowed for the creation of wealth proportionate to the electricity consumption they needed.

So, the question is not purely environmental but rests on the balance between, on one side, energy costs in money and environmental terms, and on the other side, economic and social benefits derived from the use of cryptocurrencies.

2.1.2 Systems Dynamics: Applied to the Bitcoin Case

The Digiconomist (2019) and other figures show that the energy consumption of Bitcoin mining is evolving under a number of factors that will be discussed here. The general systems dynamics is given in Fig. 1. This chart exhibits the drivers of Bitcoin mining profitability and the feedback it creates on industry players. It has been kept deliberately simple neglecting the finer points of Bitcoin mining. Several complete formal mining dynamics models have been published, notably by Prat and Walter (2018).

Bitcoin mining profitability depends on related costs and revenues. The main costs are electricity consumption needed to validate a block plus the capital expenditure, which is linked to the mining equipment bought.

On the revenue side, a miner can expect a reward when s/he is the first one to validate a block. It can also charge transaction fees—we will neglect these fees for the sake of simplicity. Revenues are therefore directly linked to the Bitcoin price: when it goes up, revenues go up.

The difference between revenues and costs gives the profit generated per new block.

Bitcoin Mining Rules

A key role is played by the Bitcoin mining rules. They make it ever more difficult and less profitable to mine through two mechanisms. First, the hashrate (measurement of mining difficulty) has been programmed to adjust over time so that one

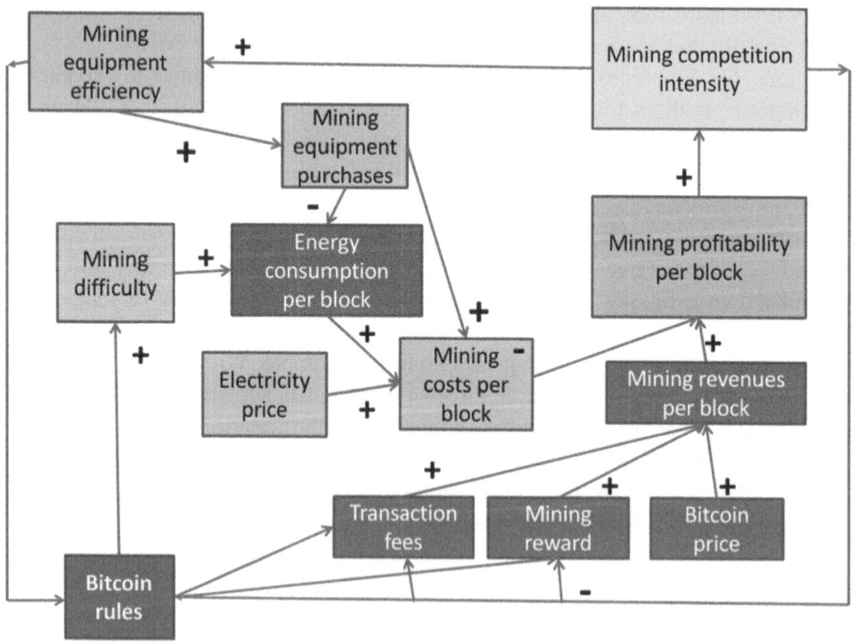

Fig. 1 General system dynamics of Bitcoin Mining

block is validated every 10 min, which means that when the computing power increases the mining difficulty increases. Then the mining reward, which is the amount of Bitcoins obtained by the miner when s/he is the first to validate a block, is programmed to decrease over time, halving every 4 years approximately. For example, it was equal to 12.5 Bitcoins/block in 2016 and expected to decrease to 6.25 Bitcoins in 2020. This mechanism mimics what happens in a real gold mine, where gold is more difficult to mine over time. One should note that the total number of Bitcoins is set at 21 million and 17 million have been mined up to May 2019: Bitcoin mining is accordingly much more difficult that what it used to be a few years back.

Bitcoin Price
On the revenue side, Bitcoin price increases have a positive impact on mining profitability as the mining reward becomes more valuable. This explains why Bitcoin mining became very popular during the 2017 price bubble.

Electricity Prices
On the other side, electricity price increases have a negative impact on mining profitability, even more so as mining difficulty is increasing.

Mining Equipment
If miners were using standard computers, mining would be too slow and very energy-inefficient, so they buy specialized computing equipment. Mining equipment advances have a positive impact on energy consumption, but they are costly to implement and generate a rebound effect. Indeed, because of the competition between miners, any saving made on each computer will tend to be spent on additional computers to gain a competitive advantage over other miners. And they have an impact on mining difficulty as explained above.

Mining Competition Intensity
Any increase in mining profitability will generate an increase in mining competition intensity in a few months as it attracts new miners or entices existing miners to enlarge their operations; but a decrease in mining profitability will only generate a decrease in mining competition intensity in the medium term. This is clearly seen on the Digiconomist bitcoin energy index, peaking when the Bitcoin price hit all time highs in late 2017 and decreasing at the end of 2018, while the Bitcoin price was much lower. It takes a few months to set up a mining farm but once it has been set up, there is not point stopping it unless the mining revenues fall below variable costs. If the mining profitability keeps below expectations, it is certain that miners will not invest again in new equipment.

So, a conclusion is that mining difficulty is ever-increasing as Bitcoin usage grows and energy consumption grows as well, albeit mitigated by advances in computing equipment and the fact that the most efficient miners are the only ones rewarded (Giungato et al. 2017). Two key issues are the behavior of miners (how much profitability they expect) and government regulation of Bitcoin mining. Indeed, we know that governments and central banks are closely watching the development of cryptocurrencies. They occasionally take measures designed to curb or promote Bitcoin mining. For example, in early 2019, according to Liao and Russell in TechCrunch (2019), China decided to include cryptomining in the list of industries to be eliminated because they "lacked safe production conditions, seriously wasted resources, polluted the environment".

2.2 *What Are the Solutions?*

Energy consumption linked to mining is therefore a serious issue. Both miners and governments look for solutions to decrease this energy consumption and several possibilities exist.

2.2.1 Green Energy

The first solution is to move mining to places where electricity is abundant, cheap and green. For example, according to Tuwiner (2019), around 70% of Bitcoin mining is conducted in China where electricity is cheap, but not always green. According to the same source, in 2019 farms can also be found in Iceland where they benefit from cheap and green electricity. However, these policies have no impact on the actual total electricity consumption level, so other solutions have to be found. The Digiconomist founder, De Vries (2019) also points out to the fact that renewable energy sources do not provide the constant supply needed by Bitcoin mining farms.

2.2.2 Mining Equipment Energy Efficiency

One way to be considered is the increased energy efficiency of mining computers. However, this is linked to the general improvement of computing technology and unlikely to provoke a big bang in mining energy efficiency.

2.2.3 System Parameters

As seen in the previous paragraph, Bitcoin rules play a major role in mining efficiency. There are many parameters in Bitcoin management that could reduce energy consumption like: the hashrate, the size of blocks, the validation method etc.

Ethereum, being a step-child of Bitcoin, has optimized many of these parameters and accordingly its electricity consumption is significantly lower than that of Bitcoin.

General Governance Rules
However, Bitcoin has been designed as a decentralized system which cannot be changed easily once the initial implementation is in place (De Filippi 2016). Changes are of course possible, but they take a long time, their outcome is uncertain and they often result in hard forks with one part of the community following the old rules and the other part following the new ones (De Filippi and Loveluck 2016). Bitcoin, however clever its design, cannot accommodate a very large number of transactions in a short period of time, in particular because of its limited block size. So, several proposals have been put forward to change this parameter. A major fork occurred in 2017 with the new Bitcoin Cash chosen by a minority, while the majority stuck with the old rules. These forking processes are very risky for Bitcoin owners as shown by the much lower price and popularity of Bitcoin Cash compared to the standard Bitcoin.

Proof of Stake Versus Proof of Work
A major improvement in the energy efficiency of mining would be to move from the historic proof of work system to another one. The obvious candidate is the proof of stake which requires very limited energy consumption as it delegates block validation to a group of stakeholders, without the need to make complicated computations.

The Ethereum community roadmap (Ethereum 2019) plans a switch to proof of stake, but it is not an easy decision to make. Security and migration issues block the process until 2019—one needs to be absolutely sure that block validation cannot be manipulated and that the validation process will run smoothly, a guarantee which has not been given at this date by the proposed variant of proof of stake.

Other validation systems have been designed, but as of 2019 none appears to be ready to replace the existing ones and anyway, governance issues would make it very difficult to change the set-up for an existing and widely used currency like Bitcoin.

2.2.4 Mining Competition

A last possibility to decrease electricity consumption would be to lessen the intensity of mining competition. With fewer miners competing to be the first to validate a block, energy consumption would be lower. In mining pools, miners share their computing power as well as the reward associated with the validation of a block. But the competition between mining pools (vital to keep the validation process running) prevents any decrease in energy consumption.

2.3 A Striking Example of Real Issues Related to Virtual Systems

The local impact of Bitcoin mining can be large in terms of energy consumption but its global socio-economic impact is "uncertain" as shown by the Greenberg and Bugden 2019 conducted on Chelan county (Washington State, USA) where "cryptomining boomtowns" have appeared.

Yet, even though it might look anecdotal, a striking figure has been provided by Krause and Tolaymat (2018) who showed that Bitcoin mining was actually more energy intensive that gold mining—a rare instance when a dematerialized procedure (for the Bitcoin) uses more energy than a traditional earth-moving technology. This is not a big surprise as, after all, Bitcoin is modeled after the gold standard.

So, the major cryptocurrencies have a real impact on energy, using resources at levels and in ways which have been considered problematic globally. Is cryptomining the "best" use of limited and polluting energy resources, while less energy-intensive monetary systems already exist? At the local level, where the mining farms are set up, there is often a fear of electricity shortages and price hikes, as well as increased pollution while local benefits (for example in terms of employment) are limited. Governments are uncertain as to the regulation of cryptocurrencies, but regulating mining farms is a possibility if miners appear to create significant problems.

3 Cryptocurrencies and the Dark Net: The Dark Side of Cryptocurrencies

Internet has long facilitated crime, but the advance in anonymity has created a radical change, mainly in two ways: through encryption activated by the hidden web (e.g. Tor) and through cryptocurrency, which masks the identity of participants' online activities (Nica et al. 2017). Indeed, with the emergence of Bitcoin in particular, a new type of Dark economy has emerged "electronic commerce on the black market" (Foley et al. 2019). Virtual money laundering and terrorist financing offer high levels of anonymity, potentially low detection, and the removal of many of the risks associated with real-world money laundering and terrorist financing activities (Irwin et al. 2014). Dark websites serve as a platform for Internet users for whom anonymity is essential, since they not only offer protection against unauthorized users but also include encryption, via transaction anonymizers, to prevent monitoring (Möser et al. 2013). Users who fear economic or political reprisals for their actions turn to the dark web to protect themselves. Nevertheless, there are also those who take advantage of this online anonymity to use the Dark Web for illegal activities, such as trading in controlled substances or illegal financial transactions (Chertoff and Simon 2015). As such, bitcoin users involved in illegal activities behave differently from other users (Foley et al. 2019). Illegal users tend to make more transactions, but with smaller amounts. They are also more likely to engage in repeated transactions with a given counterpart. Despite an increasing number of transactions, illegal users tend to hold fewer bitcoins, which is consistent with the risk of seizure of bitcoin assets by authorities.

Foley et al. (2019) find that illegal activities represent a substantial proportion (26%) of Bitcoin users and commercial activities. They estimate that about $76 billion of illegal activities per year involve Bitcoin (46% of Bitcoin transactions), which is close to the size of the US and European illicit drug market. In addition, about one-fifth (23%) of the total dollar value of transactions and about half of the Bitcoin holdings (49%) over time are associated with illegal activities. In April 2017, approximately 27 million participants in the bitcoin market used it mainly for illegal purposes.

According to Corbet et al. (2019), cybercrime can take two main forms: cybercrime resulting from the use of cryptocurrencies through the financing of illegal activities (3.1), and cybercrime affecting the direct structures of the cryptocurrencies themselves, via cyber-attacks (3.2). We have also identified in the literature a third form of cybercrime, money laundering and tax evasion (3.3).

3.1 Financing Illegal Activities

Cryptocurrencies combine the two characteristics that traditional currencies lack but which organized criminal groups search for: they provide anonymity to users and they are easy and fast to transfer worldwide and in an almost instantaneous manner. This opens up opportunities for different types of organized crime groups (Nica et al. 2017). Exchange platforms operate anonymously so that money can move secretly from one user to another, allowing for the existence of a parallel banking system or even a parallel economy for illegal products and services (Dierksmeier and Seele 2018). Moreover, law enforcement agencies are concerned that Bitcoin may provide a payment mechanism that could facilitate and increase illegal activities, such as child pornography, drug trafficking or terrorism (Angel and McCabe 2015).

3.1.1 Financing of Illicit Products

On websites specially designed to escape public scrutiny, cryptocurrencies are used to buy and sell drugs, weapons and products related to pornography, counterfeit money or stolen credit cards.

One of the best known examples of cybercrime related to the use of cryptocurrencies is the "Silk Road", which was an online black market dedicated to the sale of drugs on the dark net and on which payments were made exclusively in Cryptocurrencies (Nica et al. 2017; Corbet et al. 2019). When the website was launched in 2011, payments were made exclusively in Bitcoin, but over time, other altcoins, such as Dash and Monero, began to be used. The anonymity offered by these types of sites is valuable to people who are concerned about the confidentiality of their transactions (Angel and McCabe 2015). The FBI's seizure of more than $4 million worth of bitcoins on the "Silk Road" gives an idea of the magnitude of the problem facing regulators (Foley et al. 2019). The FBI estimated that the Silk Road had accounted for nearly 5% of the total Bitcoin economy and when the site was announced for closure, the price of Bitcoin fell from $145 to $109.

According to Foley et al. (2019), there is no doubt that, through a digital and anonymous payment mechanism, cryptocurrencies such as Bitcoin have facilitated the growth of online "dark net" markets in which illicit goods and services are traded. Although law enforcement agencies have recently been successful in closing several e-commerce sites, it seems undeniable that the advent of cryptocurrencies has changed the playing field in favor of criminally motivated merchants and consumers (Dierksmeier and Seele 2018). It has already been proven that the increasing revenue opportunities and reduced probability of detection attract an ever-increasing number of illegal goods.

However, since 2016, the proportion of bitcoin activities associated with illegal trade has decreased, although the absolute amount has continued to increase. Foley et al. (2019) attribute this decrease to two main factors. The first is the rapid growth of the speculative interest in Bitcoin, which mechanically reduces the illegal share. The second factor is the emergence of alternative cryptocurrencies, which are more opaque and capable of further concealing a user's activity (for example, Dash or Monero). Nevertheless, despite the emergence of alternative cryptocurrencies and numerous seizures made in dark markets by law enforcement agencies, the number of illegal activities involving Bitcoin remains close to its record level in April 2017 (Foley et al. 2019).

3.1.2 Terrorist Financing

Cryptocurrencies have also been linked to terrorist financing cases (Durrant 2018). In general, terrorists are financed by private donations, non-profit organizations, criminal activities, extortion of local populations, kidnappings for ransom, etc. The Financial Action Task Force (FATF) report (2015) indicates that terrorists then use physical methods, i.e. personal letters, and virtual methods, i.e. cryptocurrencies, to repatriate funds to the organizations to which they belong. Irwin and Milad (2016) clearly establish the risks posed by cryptocurrencies and more particularly Bitcoins to facilitate the process of financing, planning and implementing terrorist acts, even if the extent of the phenomenon is difficult to assess. However, there is much evidence to suggest that they have been linked to many terrorist attacks in Europe and Indonesia.

With better access to the Internet, particularly to promote propaganda, terrorists use cryptocurrencies to launder the money they receive from crimes, but also as a means for foreign donors to support them financially without reprisals from their home countries (Durrant 2018). For example, supporters of ISIS, jihadists and terrorist organizations have been identified in forums, websites and social networks, and have used social networks to ask people to finance jihad with Bitcoin. Deutsche

Welle reported that a portfolio of bitcoins, which appears to belong to ISIS, received $23 million in payments over 1 month (Nica et al. 2017). Irwin and Milad (2016) suspect that one of the simplest supposed methods of exchanging bitcoins for cash is two-way ATMs, which can be purchased by the organization itself and used only to transfer money between its units in an international, anonymous and almost instantaneous manner. Many Bitcoin ATMs are located in countries that have seen a significant number of fighters join the Islamic State group in the Middle East and are located in countries where the risk of terrorist attack is increased.

In conclusion, even if the financing of terrorism by cryptocurrencies does not occur as radically as some law enforcement authorities suggest (or perhaps supporters of terrorism are skilled at hiding their identity), this type of financing does exist (Durrant 2018).

3.2 *Cyber Attacks*

Another form of cybercrime, cyberattacks, affects cryptocurrencies (Corbet et al. 2019). Attacks on cryptocurrencies can be a real and significant threat (Nica et al. 2017). Moreover, Dierksmeier and Seele (2018) even assert that the major risk for individuals using cryptocurrencies comes from hackers who "steal" altcoins through unauthorized access to their electronic wallets or digital exchange platforms. It is estimated that pirates steal nearly 10% of all ICO revenues. Although this is an incredible indictment of the ICO process, it is not the only mechanism by which investors in cryptocurrencies have been swindled. Indeed, piracy of stock exchanges and cryptocurrencies portfolios has spread and worsened in recent years. For example, Bitfinex is a Hong Kong-based cryptocurrency exchange platform owned by iFinex Inc. founded in 2010; Bitfinex has quickly reached the top of the Bitcoin trading market. It collapsed after the largest hacking took place on the platform in 2014, resulting in the theft of more than 700,000 bitcoins worth about $473 million (Corbet et al. 2019). While many advocates of cryptocurrencies see the absence of a regulatory body as a reason to have more confidence in cryptocurrencies than in real currencies, this approach also makes them powerless in the event of piracy (Nica et al. 2017). The current lack of deposit insurance prevents users of Bitcoin platforms from facing a system crash or currency theft. However, this problem could be solved by the emergence of both the private insurance and public deposit insurance industries and by regulatory efforts similar to provisions in the world of traditional currencies (Dierksmeier and Seele 2018).

In addition to the theft of altcoin portfolios, cryptocurrencies have also facilitated the spread of ransomware attacks, such as WannaCry (Nica et al. 2017). Ransomware is malicious software that blocks companies' access to their own data and only unlocks them against the payment of a ransom in cryptocurrencies. The Armada's collective attack on Greek banks in 2015 (Brown 2016) and WannaCry's attack on many global companies and organizations, are other examples of ransomware that used bitcoins as the preferred payment method. Ransomware has become one of the most profitable attacks in history, and the results of the UK National Crime Agency's survey indicate that some companies are now storing bitcoins in anticipation of a ransomware attack (Levin et al. 2015). A Google study found that ransomware victims had paid more than $25 million in ransoms in the last 2 years (Brandom 2017).

3.3 Money Laundering and Tax Evasion

Digital currencies" and "virtual worlds" offer criminals opportunities for money laundering because of their global reach, the absence of face-to-face transactions and the convenience of electronic commerce. Although the nature and extent of money laundering through digital currencies and virtual worlds is unknown, it is important to recognize their potential for criminal exploitation (AUSTRAC 2012, p. 8).

By definition, money laundering exploits the vulnerabilities of products and services to conceal the proceeds of illicit activities and to commit financial and other serious crimes. Money laundering is also inherent in serious tax evasion (AUSTRAC 2012, p. 4). Typically, money laundering involves three steps to conceal the source of illicit funds and give them a legitimate appearance (AUSTRAC 2014). First, placement is where illicit funds are introduced into the formal financial system by depositing small amounts of cash into different bank accounts. Second, layering is money "washing" or dispersion through several transactions to hide its true origin (e.g. use of a series of complex transactions involving several banks and/or companies). Finally, integration of money into the legitimate circulation via the investment of funds—now distanced—into other legitimate business activities or the purchase of high-value assets and luxury products. Cryptocurrencies have already been identified as "potentially vulnerable" to money laundering by AUSTRAC (2012). In the "Potential vulnerabilities" section of its 2012 report, it examines a number of channels vulnerable to money laundering and terrorist financing, including digital currencies and virtual worlds. They can be used at each of the three stages of the cycle (AUSTRAC 2014; Nica et al. 2017).

For example, money laundering via bitcoins can be done using mixing portfolios that transfer the bitcoins via a network of false transactions that are much more difficult to track and which increase the anonymity of the money transfer. With regard to the integration stage, online gaming services and the purchase of tokens are the most common methods used. The task of linking a pseudonym to a real person is generally impossible, making cryptocurrencies a "safe" way to launder money for criminals.

There are also institutions, which, due to the lack of regulation, can be used for money laundering and tax evasion. An example of such an institution is "Liberty Reserves", in which operators of a global exchange system have put a money laundering operation online. This platform has gone beyond the traditional limits of US and international banking regulation in what prosecutors have described as dark cyber-financing (Dierksmeier and Seele 2018). It traded in virtual currency and provided an anonymous and easily accessible banking infrastructure to such an extent that Richard Weber, who heads the Internal Revenue Service's criminal investigation division in Washington, stated that"[i]f Al Capone were alive today, that is how he would hide his money" (New York Times 2013). The site is estimated to have laundered more than $6 billion between 2006 and 2013 (Santora et al. 2013).

There is therefore a clear consensus that virtual environments and currencies pose a threat to money laundering and terrorist financing. What is less clear, however, is the level of risk they pose (Irwin et al. 2014). Indeed, some believe that virtual environments and virtual currencies do not provide the scale necessary for large-scale money laundering activity (AUSTRAC 2012). As a rule, digital currencies are not commonly accepted for the payment of regular goods and services. This limits the possibilities of using digital currency to convert, move and launder illicit funds. The limited size of digital currency markets, in turn, reduces the possibility of transferring large quantities of illicit value. The overall utility of digital currencies for criminals at this stage can be limited to niche crimes in the cyber-environment and illicit activities, individual or on a smaller scale (AUSTRAC 2012, p. 9). Irwin et al. (2014) also believe that the high levels of anonymity offered by cryptocurrencies are detrimental to ease, time and, in some cases, the amount of laundered funds. Consequently, while it is theoretically true that large sums (millions of dollars) can be laundered in virtual environments, in practice this exponentially increases the level of effort involved in the creation, layering and integration of funds. Nevertheless, the speed with which money launderers and/or terrorists have turned to virtual environments and digital currencies, when traditional sources of funding are restricted or lost, has become a challenge for governments (e.g. counter-terrorism agencies) and security professionals (AUSTRAC 2012).

4 Ethical Aspects and Uses of Cryptocurrencies

This section deals with the ethical and moral aspects of cryptocurrencies.

Cryptocurrencies can provide solutions to societal problems as reported by Vigna and Casey (2015) and Dopfer et al. (2004). In this way, they can help reduce poverty, transaction costs, debt crisis and hyperinflation.

4.1 Business Ethics Beyond Government Control

Business ethics and the impact of crypto money on trust in commercial relationships have been discussed in the literature (Dierksmeier and Seele 2018). Regarding the contribution of crypto currencies to business ethics, Dierksmeier and Seele (2018) indicate that this currency does not require the intermediation of a trusted third party such as a bank or clearing house to ensure the transfer of funds between payer and receiver with no personal ties. The absence of an intermediary means that transaction costs are low, thus facilitating micro-payments without a minimum transaction amount (Nian and Chuen 2015). In addition, the crypto currency is exchanged on a peer-to-peer computer network so that participants are directly connected to each other via the Internet, without any institutional controller (Lemieux 2013). In Bitcoin, anyone can become a miner, it simply represents a computer connected to the Internet that performs the necessary calculations to verify each transaction (Angel and McCabe 2015). The advantage of this mechanism is that each user can enter and leave the network at any time. In this context, the transaction is guaranteed by an impersonal verification technology that also ensures its transparency. This helps to prevent fraud in the sense that this technology verifies that the transfer of a given amount of value has been made from one party to another. This process allows users of virtual currencies to trust a diverse range of profit-driven miners (Angel and McCabe 2015). This system therefore seems very secure, unlike credit card transactions, which can use magnetic strips and unsecured signatures (Nian and Chuen 2015). The irreversibility of the transaction also protects and guarantees the income of merchants. The latter may have their transaction cancelled in the event of a dispute with the buyer during transactions carried out in a traditional manner, unlike those carried out via a cryptocurrency (Mas and Chuen 2015). However, care should be taken in this regard, as in the event of proven fraud, the payer/consumer cannot be protected and his funds will be permanently lost (Mas and Chuen 2015).

4.2 Reduction in Inflation and Central Bank Adjustments

The absence of institutional control implies that, unlike real currencies, this digital currency cannot be manipulated or modified (Lemieux 2013). Indeed, central banks can regulate the volume of traditional money in the market through tools ranging from interest rate adjustments to quantitative easing (Kleineberg and Helbing 2016). These mechanisms are intended to stimulate the economy and control inflation. However, these actions do not always lead to the desired objective, which raises concerns about the central bank's ability to act.

Hence, a cryptocurrency avoids inflation and places virtual money as a potential safe haven (Dierksmeier and Seele 2018). Thus, at the macroeconomic level, a crypto-currency can lead to a more stable, sustainable and equitable economy (Maurer et al. 2013; Angel and McCabe 2015).

4.3 Poverty Reduction

Moreover, in the digital age and in the absence of financial intermediation, the various transactions can be carried out by everyone in a dematerialized, easy and instantaneous way, without geographical barriers. This leads to low transaction costs compared to traditional means of transferring funds or paying via Visa and MasterCard systems (Angel and McCabe 2015). To this end, these low costs—ranging from 0 to 1% of the amount—have a considerable impact on donations or transactions of a human and social nature, such as assistance to victims of natural disasters, international remittances from migrants to their families in developing countries, etc. (Dierksmeier and Seele 2018; Nian and Chuen 2015; Angel and McCabe 2015). This system reduces barriers to remittances and allows beneficiaries to benefit from the amount transferred at low cost. Angel and McCabe (2015) indicate that the receiver of a Bitcoin payment costs almost nothing, even if this merchant has costs related to the installation of the software, which manages the bitcoins, the transaction costs related to the conversion of the bitcoins into other currencies, as well as the risk of handling several types of currencies. As a result, if you have young entrepreneurs who are viable thanks to many small payments or donations from contributors, these low transaction costs can have a positive impact on their bottom line and business model. This digital currency also helps to overcome the difficulties faced by some individuals who cannot open a traditional bank account and thus allows them to take part in transactions via the Internet. It should be noted that some individuals do have constraints when trying to open a regular bank account because of their banking history, inappropriate home address, irregular situation, etc. These individuals have the ability to bypass these problems by turning to virtual currencies.

These advantages make cryptocurrency an interesting alternative to other forms of payment. This is all the more noticeable in developing countries where mobile payments and money transfers by mobile phone are becoming more popular (Dierksmeier and Seele 2018).

4.4 Protection of Privacy and Personal Data

One of the driving forces of cryptocurrencies is the protection of privacy and anonymity of transactions (Maurer et al. 2013). This system is suitable for people who wish to guarantee an appropriate level of protection of their personal data. The owner or user has a private key to carry out transactions and to use his wallet, he will not be surprised by the additional costs and even less by the theft of his bank details as can be the case when credit cards are lost or misused. This protection is due to the fact that transactions do not contain much personal information (Nian and Chuen 2015). In addition, the identity of the user of digital currencies is not required, as well as other information often required by the bank (e.g. identity card, postal address, etc.), which greatly reduces the theft of personal data (Mas and Chuen 2015). However, in view of the theft of digital currency portfolios, users must take precautions just like other financial products to protect themselves against these risks of loss or theft (Nian and Chuen 2015).

5 Conclusion: Regulatory Issues and Challenges

Cryptocurrencies have many potential benefits, including faster, more efficient and a less costly payment settlement which facilitates micro-payments without a minimum transaction amount, the reduction of poverty, the prevention of hyperinflation. However, the regulatory concerns focus on their use in illegal trade (drugs, piracy and theft, illegal pornography), cyber-attacks, the potential for terrorist financing, money laundering, and tax evasion (Foley et al. 2019; Irwin et al. 2014). Cryptocurrencies have been identified in the National Strategic Assessment of Serious and Organized Crime 2017 as technological threats that pose a particular challenge to authorities (Nica et al. 2017). Moreover, the extreme volatility that cryptocurrencies can experience is often associated with their lack of regulation and associated cybercrime (Corbet et al. 2019). According to Foley et al. (2019), a significant part of the intrinsic value of bitcoin as a payment system is even derived from its propensity to be used to facilitate illegal trade. This has ethical implications for those who consider bitcoin as an investment, in addition to the implications in terms of valuation. For example, changes in the demand for the use of bitcoin in illegal trade (because of law enforcement repression or the increasing adoption of more opaque cryptocurrencies in illegal trade) are likely to affect its fundamental value. A decrease in the illegal share of bitcoin activity with the emergence of new cryptocurrencies has already been observed.

Although cryptocurrencies have disadvantages, critics argue that the innovation of Bitcoin technology that promotes a free global market and financially connects the world is worth the risk, as it helps many more people than it hurts (Durrant 2018). Moreover, it is generally accepted that if cryptocurrencies attract cybercriminals it is because they have two main advantages over transactions in the real world: anonymity and low costs. However, the scope and extent of these two advantages are questioned in the literature. Regarding anonymity, for example, Irwin et al. (2014) argue that real currencies can only remove a number of risks and reduce the chances of detection against real currencies if the money launderer or terrorist takes drastic measures to eliminate traces of his or her true electronic identity and location. Otherwise, he would be detectable and traceable, as well as the accounts he has opened and the transactions he has carried out. Recently, there has been increasing success in digital tracing and thus in controlling illegal trade in the blockchain world by national or international law enforcement authorities (Dierksmeier and Seele 2018). Regarding costs, it would seem that cryptocurrencies pose significant risks for their "honest" (risk of cyber-attacks) and "malicious" (risk of detection and seizure of their portfolios) users, without really providing a reduction of costs (Irwin et al. 2014; Nica et al. 2017).

However, it remains difficult to balance the promotion of innovative technology while deterring the associated crime. This again raises the old ethical question of how to balance the abuse potential of a product with its benefits. As long as a product has significant potential benefits, an ethical judgment should be made about the use of the product, not about the product itself (Angel and McCabe 2015). For example, it would be interesting to know that, like Bitcoin, the foundations of Silk Road were based on the libertarian ideology "which allowed users to buy and sell property, although illegal, if it does not interfere with or destroy the ownership of others' property or their physical integrity", although it is legitimate to wonder whether one can really consider that selling drugs to people does not necessarily destroy their physical integrity (Nica et al. 2017). For this reason, Silk Road and many other dark markets have refused to sell products used to harm others, such as weapons of mass destruction or child pornography. The movement and its users refer to "dark ethical markets" that only offer services that are unavailable or costly due to government regulation, but also to those that do not interfere with the rights of others, namely those that are penalized, although they are victimless crimes. In order to satisfy their customers' demand, some of these dark markets have begun to offer drugs that they claim to be "fair trade" or manufactured in areas not affected by war.

The unestablished legal status of cryptocurrencies has also benefited those who use cryptocurrencies for illegal activities and has prevented rapid adaptation by merchants and other legal businesses. The European Central Bank does not recognize virtual currencies as a currency (ECB 2015), either economically or legally, and at present, the EU does not regulate virtual currencies, which are therefore not subject to the Payment Services Directives or Earnest Money Deposit regulations. Due to their decentralized nature, monetary systems such as Bitcoin do not have a centralized entity capable of monitoring and reporting suspicious activities (Nica et al. 2017). In 2015, the FATF published a report specifically on cryptocurrencies with the aim of establishing a conceptual framework to combat the risks associated with the system for combating money laundering and terrorist financing. The European Commission and EUROPOL also discussed the subject of cryptocurrencies. In 2014, the Canadian Parliament passed a bill amending the current anti-money laundering and anti-terrorist financing laws to apply to persons using virtual currencies.

As an emerging innovation in the financial technology sector, cryptocurrencies and the blockchain technology on which they are based could revolutionize many aspects of the financial system, from smart contracts to settlements, from interbank transfers to venture capital funds, as well as applications beyond the financial system. Like many innovations, cryptocurrencies also have a dark side. We have clarified this dark side by describing their use in illegal activities. We also discussed their environmental impact. Dernbach and Brown (2009) argue that it is our ethical responsibility to reduce energy consumption, but all human activities use energy, so the question, once again rests on the balance between costs and benefits.

Finally, we recognize that the field of research related to cryptocurrencies is immature and that empirical and theoretical evidence continues to emerge every month. Although cryptocurrencies continue to develop both as a product and as a negotiated market, it is important that expectations about their potential value and benefits to society are moderate, while remaining cautious and considering the inherent dangers they could generate for society (Angel and McCabe 2015).

References

Aldridge J, Décary-Hétu D (2014) Not an 'Ebay for Drugs': the Cryptomarket 'Silk Road' as a paradigm shifting criminal innovation. Available via SSRN 2436643. Accessed 29 Apr 2019

Angel JJ, McCabe D (2015) The ethics of payments: paper, plastic or Bitcoin? J Bus Ethics 132 (3):603–611

Australian Transactions Reports and Analysis Center (AUSTRAC) (2012) Typologies and case studies report. Australian Government. http://www.austrac.gov.au/typologies-and-case-studies-report-2012. Accessed 23 Apr 2019

Australian Transactions Reports and Analysis Center (AUSTRAC) (2014) Typologies and case studies report. Australian Government. http://www.austrac.gov.au/sites/default/files/typologies-report-2014.pdf. Accessed 16 May 2019

Beigel O (2018) What is Darkcoin|How to buy darkcoin? https://99bitcoins.com/darkcoin-buy-darkcoin/. Accessed 19 Apr 2019

Bevand M (2017) Electricity consumption of Bitcoin: a market-based and technical analysis. http://blog.zorinaq.com/bitcoin-electricity-consumption/ Accessed 10 March 2017

Blockchain Luxemburg SA (2019) https://www.blockchain.com/charts/blocks-size

Brandom R (2017) Ransomware victims have paid out more than $25 million, Google study finds. July 25. https://www.theverge.com/2017/7/25/16023920/ransomware-statistics-locky-erbergoogle-research

Brown SD (2016) Cryptocurrency and criminality: the Bitcoin opportunity. Police J 89(4):327–339

Chertoff M, Simon T (2015) The impact of the dark web on internet governance and cyber security, Global Commission on Internet Governance. In: Paper series n°6, the royal institute of International affairs. Waterloom Ontario : CIGI Chatham House, Ottawa, ON. 59p. https://www.cigionline.org/sites/default/files/gcig_paper_no6.pdf. Accessed 29 Apr 2019

Corbet S, Lucey B, Urquhart A, Yarovaya L (2019) Cryptocurrencies as a financial asset: a systematic analysis. Int Rev Financ Anal 62:182–199

De Filippi P (2016) The interplay between decentralization and privacy: the case of blockchain technologies. J Peer Prod 7

De Filippi P, Loveluck B (2016) The invisible politics of Bitcoin: governance crisis of a decentralized infrastructure. Internet Policy Rev 5:4

De Vries A (2019) Renewable energy will not solve Bitcoin's sustainability problem. Joule 3 (4):893–898. https://doi.org/10.1016/j.joule.2019.02.007

Dernbach JC, Brown DA (2009) The ethical responsibility to reduce energy consumption. Hofstra L Rev 37:985

Dierksmeier C, Seele P (2018) Cryptocurrencies and business ethics. J Bus Ethics 152(1):1–14. https://doi.org/10.1007/s10551-016-3298-0

Digiconomist (2019) https://digiconomist.net/bitcoin-energy-consumption

Dilek Ş, Furuncu Y (2019) Bitcoin mining and its environmental effects. Ataturk Univ J Econ & Adm Sci 33(1):91–106. http://dergipark.org.tr/atauniiibd/issue/43125/423056

Dopfer K, Foster J, Potts J (2004) Micro-meso-macro. J Evol Econ 14(3):263–279

Durrant S (2018) Understanding the nexus between cryptocurrencies and transnational crime operations. CUNY Academic Works, New York. https://academicworks.cuny.edu/jj_etds/70/

Dyhrberg AH (2016) Bitcoin, gold and the dollar–A GARCH volatility analysis. Financ Res Lett 16:85–92

Ethereum 2.0 (Serenity) Phases. (2019). https://docs.ethhub.io/ethereum-roadmap/ethereum-2.0/eth-2.0-phases/. Accessed 14 May 2019

European Central Bank (ECB) (2015) Virtual currency schemes – a further analysis. European Central Bank, Staff Working Paper

Financial Action Task Force (FATF) (2015) Emerging terrorist financing risks. FATF, Paris. http://www.fatf-gafi.org/media/fatf/documents/reports/Emerging-Terrorist-FinancingRisks.pdf

Foley S, Karlsen JR, Putniņš TJ (2019) Sex, drugs, and Bitcoin: how much illegal activity is financed through cryptocurrencies? Rev Financ Stud 32(5):1798–1853

Giungato P, Rana R, Tarabella A, Tricase C (2017) Current trends in sustainability of Bitcoins and related blockchain technology. Sustainability 9(12):2214

González PRV (2016) Superabundant design: from waste to control in Bitcoin mining. A peer-reviewed Journal About 5.1. http://www.aprja.net/superabundant-design-from-waste-to-control-in-bitcoin-mining/

Greenberg P, Bugden D (2019) Energy consumption boomtowns in the United States: community responses to a cryptocurrency boom. Energy Res Soc Sci 50:162–167. https://doi.org/10.1016/j.erss.2018.12.005

Guesmi K, Saadi S, Abid I, Ftiti Z (2018) Portfolio diversification with virtual currency: evidence from Bitcoin. Int Rev Financ Anal 63:431–437. https://doi.org/10.1016/j.irfa.2018.03.004

Hayes A (2015a) A cost of production model for Bitcoin. Available via SSRN 2580904. https://papers.ssrn.com/sol3/papers.cfm?abstract_id=2580904

Hayes A (2015b) The decision to produce altcoins: miners' arbitrage in cryptocurrency markets. March 16, 2015. Available via SSRN 2580904. https://papers.ssrn.com/sol3/papers.cfm?abstract_id=2579448

International Money Fund Staff Team (2016) Virtual currencies and beyond: initial considerations. IMF staff discussion Note. https://www.imf.org/external/pubs/ft/sdn/2016/sdn1603.pdf. Accessed 30 Apr 2019

Irwin ASM, Slay J, Raymond Choo KK, Lui L (2014) Money laundering and terrorism financing in virtual environments: a feasibility study. J Money Laund Control 17(1):50–75

Irwin ASM, Milad G (2016) The use of crypto-currencies in funding violent jihad. J Money Laund Control 19(4): 407–425

Kleineberg KK, Helbing D (2016) A "Social Bitcoin" could sustain a democratic digital world. Eur Phys J Spec Top 225(17–18):3231–3241

Krause MJ, Tolaymat T (2018) Quantification of energy and carbon costs for mining cryptocurrencies. Nat Sustain 1(11):711. https://doi.org/10.1038/s41893-018-0152-7

Krugman P (2013) Bitcoin is evil. https://krugman.blogs.nytimes.com/2013/12/28/bitcoin-is-evil/. Accessed 17 Apr 2019

Lemieux P (2013) Who is Satoshi Nakamoto? Regulation 36(3):14–16. Academic One File, Accessed 9 May 2019

Levin RB, O'Brien AA, Zuberi MM (2015) Real regulation of virtual currencies. In: Chuen DLK (ed) Handbook of digital currency. Elsevier, Academic Press, pp 327–360. https://doi.org/10.1016/B978-0-12-802117-0.00017-5

Liao R, Russell J (2019) Regulators in China are weighing a ban on Bitcoin mining Dent Tech 9 April 2019 https://techcrunch.com/2019/04/09/china-considers-ban-crypto-mining/

Mas I, Chuen DLK (2015) Bitcoin-like protocols and innovations. In: Handbook of digital currency. Elsevier, Academic Press, pp 417–451. https://doi.org/10.1016/B978-0-12-802117-0.00021-7

Maurer B, Nelms TC, Swartz L (2013) "When perhaps the real problem is money itself!" the practical materiality of Bitcoin. Soc Semiot 23(2):261–277. https://doi.org/10.1080/10350330.2013.777594

Mora C, Rollins RL, Taladay K, Kantar MB, Chock MK, Shimada M, Franklin EC (2018) Bitcoin emissions alone could push global warming above 2° C. Nat Clim Chang 8(11):931. https://doi.org/10.1038/s41558-018-0321-8

Möser M, Böhme R, Breuker D (2013, September) An inquiry into money laundering tools in the Bitcoin ecosystem. In: 2013 APWG eCrime Researchers Summit. IEEE, pp 1–14. https://doi.org/10.1109/eCRS.2013.6805780

Nian LP, Chuen DLK (2015) Introduction to Bitcoin. In: Chuen DLK (ed) Handbook of digital currency. Elsevier, Academic Press, pp 5–30. https://doi.org/10.1016/B978-0-12-802117-0.00001-1

Nica O, Piotrowska K, Schenk-Hoppé K R (2017) Cryptocurrencies: economic benefits and risks. University of Manchester, FinTech working paper, (2). https://papers.ssrn.com/sol3/papers.cfm?abstract_id=3059856

O'Dwyer K J, Malone D (2014). Bitcoin mining and its energy footprint." ISSC 2014/CHCT 2014, Limerick, June 26–27. 280–285, doi: https://doi.org/10.1049/cp.2014.0699
Plassaras NA (2013) Regulating digital currencies: bringing Bitcoin within the reach of IMF. Chic J Int Law 14(1):377–407
Prat J, Walter B (2018) An equilibrium model of the market for Bitcoin mining. CESifo Working Paper Series No. 6865. https://papers.ssrn.com/sol3/papers.cfm?abstract_id=3143410
Santora M, Rashbaum WK, Perlroth N (2013) Online currency exchange accused of laundering $6 billion. New York Times, 28/05/2013 https://www.nytimes.com/2013/05/29/nyregion/liberty-reserve-operators-accused-of-money-laundering.html. Accessed 26 Apr 2019
Sovacool BK, Sidortsov RV, Jones BR (2013) Energy security, equality and justice. Routledge, London. https://doi.org/10.4324/9780203066348
Tuwiner J (2019) Bitcoin mining in China. https://www.buybitcoinworldwide.com/mining/china/, 28/01/2019. Accessed 18 Oct 2019
Urquhart A, Zhang H (2018) Is Bitcoin a hedge or safe-haven for currencies? An intraday analysis An Intraday Analysis (January 31). https://papers.ssrn.com/sol3/papers.cfm?abstract_id=3114108
Vigna P, Casey MJ (2015) The age of cryptocurrency: how Bitcoin and digital money are challenging the global economic order. St. Martin's Press, New York
Wimbush S (2018) Letter to nature. Nature, 22 March 2018, 555: 443

Cryptocurrency Mining

Vikrant Gandotra, François-Éric Racicot, and Alireza Rahimzadeh

Abstract This chapter is mainly concerned with outlining the process of cryptocurrency mining. There are dozens of altcoins which you could mine; however, Bitcoin by far is the most popular choice for cryptocurrency miners. This is primarily due to the advancement in technology concerning the hardware and software used to mine cryptocurrencies. Specialized hardware and software are now being designed for the sole purpose of mining Bitcoin. Miners have a variety of hardware and software to choose from depending on their mining strategy. This chapter introduces the basic concept of mining, its essential functions, the hardware and software required for mining, different methods of mining and the factors influencing mining.

1 Introduction

It is a well-known fact that cryptocurrency is not issued or controlled by central authorities like governments or banks; instead, it is created from scratch. The process of creating a cryptocurrency is referred to as mining. In its most basic form, the term mining refers to how we can calculate the value of cryptocurrency assets through cryptographic processes. These processes generally mine the cryptocurrency of interest into blocks which are nothing but simplified ledger files that have a record of all recent transactions. Mining is one of the most critical aspects of any cryptocurrency protocol and is generally considered to be quite expensive and time-consuming (Malone and O'Dwyer 2014). Mining as a process enables a cryptocurrency transaction to take place, i.e., it establishes transaction capability, transaction legitimacy, and transaction consensus. Mining is quite complex to understand due to it being a mix of different disciplines, to truly understand the process, it is necessary to combine knowledge from economics, computer science, and cryptography (Berentsen and Schar 2018). The mining process can be thought of

V. Gandotra · F.-É. Racicot (✉) · A. Rahimzadeh
Telfer School of Management, Ottawa, ON, Canada
e-mail: vgand095@uottawa.ca; telfer.Racicot@uottawa.ca; arahi086@uottawa.ca

S. Goutte et al. (eds.), *Cryptofinance and Mechanisms of Exchange*, Contributions to Management Science, https://doi.org/10.1007/978-3-030-30738-7_3

as an interaction between technology and capital, and it highlights the inseparability of the two.

In the Bitcoin protocol, transactions are organized into blocks and are then added to the ledger by linking with the previously established blocks. This process of linking valid transactions together in blocks is commonly referred to as the blockchain. The information concerning these blocks and all the transactions in the blocks are stored in the disk storage of the user which is referred to as a node. These nodes store all of the information concerning the recorded transactions of the network and check the validity and legitimacy of new transactions made by using the previous blocks. The nodes are then rewarded after a validity check of the new transactions. This process is called mining (Yli-Huumo et al. 2016). This whole process then goes through another check referred to as Proof-of-Work (POW), which is one of the most critical aspects of blockchain technology. Mining as a process enables to keep the network secured by verifying the transactions that have taken place before recording them into the ledger. As mentioned earlier mining can be said to have three essential functions—creating a new currency from scratch, verifying the legitimacy of transactions and adding transactions to the block. When all transactions are successfully validated and confirmed, a concurrence occurs between the nodes. This concurrence leads to the new blocks being linked to the old blocks thereby forming a blockchain.

2 How Does Mining Work?

Now that we have explained the underlying meaning of mining, the next question arises—how does it work? Cryptocurrency mining is a complicated process and needs much investment in the form of effort, time, money, computational power and most of all energy consumption. The mining process is of critical importance as it enables the users to take part in transactions such as sending and receiving, and it enables the transaction to be verified. Whenever we mine a cryptocurrency whether it is Bitcoin, Ethereum or any other popular cryptocurrency, we are basically solving a complicated math problem/equation (Mining puzzle). The network provides this math problem at the time of transaction and when the problem is solved the transaction is approved. Computers/Mining Rigs that are utilized to mine the cryptocurrency generally utilize the blocks of transactions and convert them into complex math problems. Now, in order to solve these math problems or blocks, a large amount of computational power is required. To solve these blocks, miners use the transaction data and use unique hash functions to reach a solution. A hash value can be referred to as a unique identifier for the transaction data; it is a series of number that can be used to identify the transaction data (Nakamoto 2008).

A number of different cryptocurrencies use the SHA-256 (Secure Hash Algorithm) hash function which is a cryptographic hash function standardized in 2001. It comprises of 256 bits of state split into eight 32-bit words which makes it compatible with 32-bit hardware. The main aim of the miners is to compute these unique values to solve the block. Miners utilize massive amount of computational power to find these hash values and the one to find this value first by solving the block is the one who is said to mine the block and eventually the one who gets rewarded. In the context of Bitcoin, the reward for solving or mining a block is 12.5 BTC.[1] The information contained in these blocks is used by the miners to create the hash value. This unique hash value is then used to validate the information. In the case of any information being changed in the block, the value of the hash also changes; hence, the unique hash value is critical in maintaining the security of the network. Solving and creating this unique hash value is what generates the Bitcoin, miners tend to compete with each other to try and solve as many hashes as quickly as they can in order to mine more Bitcoin. As the use of massive amount of computational power is quite common to create these hash values, blockchain uses the POW to check if the hash values created are indeed valid and every block that gets added to the blockchain has to go through this process of validation (Information systems security 2015). When mining Bitcoin, the SHA-256 POW system is the most commonly used. The SHA-256 transforms the input message into a 256-bit message input. This Proof-of-Work is attached to every individual block so that it can be verified and added to the blockchain. Furthermore, as mentioned earlier, each block has to go through a validation process which is based on the unique hash value and altering a block requires reworking on all the following blocks. Each block in the blockchain confirms the integrity of the previous block all the way back to the initial block referred to as the genesis block. This process keeps the blockchain from being tampered with and ensures the security of the blockchain (Cocco and Marchesi 2016).

2.1 *Methods of Mining*

It is essential to differentiate between the way an individual or an organization can go about mining cryptocurrency as different methods of mining have different costs and factors that need to be considered before deciding on the most appropriate method. The three most common ways to mine are solo mining, mining contracts, and mining pools (Bhaskar and Chuen 2015).

- Solo Mining—It refers to the way in which miners solve the unique hash value individually, and the reward for solving the unique hash value is paid entirely to the individual who owns the computing power. It would seem that solo mining would be a perfect fit for mining any cryptocurrency; however, it is far from the truth. As the mining process is random and memory-less, the chance of mining any cryptocurrency are quite low and vary massively. A well-equipped individual miner may take several months to solve a block by generating the hash value. There is also a possibility that a miner may go months without solving a block due to the limited computational power, and as new blocks are added the difficulty to

[1]The BTC reward halves every 210,000 blocks, on the 23rd of May 2020, the reward will decrease to 6.25 BTC.

solve forthcoming blocks generally increases, i.e., the mining difficulty changes after every 2016 blocks. The difficulty is based on the time taken to solve the previous 2016 blocks; each block takes approximately 10 min to solve; hence, 2016 blocks would take exactly 2 weeks to solve. However, if the previous 2016 blocks took longer than 2 weeks, then the difficulty is reduced. To put this into perspective let us look at an example, the average time taken to solve a block can be given by the following calculation.[2]

$$\text{Time} = \frac{\text{Difficulty} \times \left(2^{32}\right)}{\text{Hashrate}}$$

Where difficulty is a measure of how difficult it is to compute the unique hash value and hash rate is the number of hashes the miner computes in a second. Assuming the hash rate is 1 GH/s (1 billion hashes per second), the miner uses the computational power 365 days a year for mining and the current difficulty level of 6,379,265,451,411. Then the average time for a solo miner to solve the block would be 868,808 years (approx.). As you can see this not the most ideal method for mining cryptocurrency.

- Cloud Mining Contracts—Another way to get involved in mining cryptocurrency is by utilizing cloud mining contracts. Mining contracts are ideal for those who do not want the unnecessary headache of managing and investing in hardware and software. A user only needs a standard computer for communication purposes and a wallet to store the cryptocurrency. Mining contracts as the name suggests, provide performance contracts for mining for a specific duration of time. Contracts vary from hourly to yearly duration and the main factor to consider is the difficulty level of mining. The difficulty level is a significant factor that determines the profitability of such contracts. Mining contracts can be further divided into three subcategories

  - Virtual Hosted Machines—This is perhaps the most commonly used model for cloud mining. It involves miners setting up their own virtual private servers with customized mining software installed.
  - Hosted Mining—This model is more feasible for miners with some experience and a degree of expertise in cryptocurrency mining. This model involves leasing a machine owned by a cloud mining service provider. The miner can store the leased device with the company, or have it shipped to them directly if preferred.
  - Leased Hashing Power—Another standard model of mining is to lease the hashing power. In this model, miners can buy mining contracts from a service provider without investing in any hardware or any sort of physical device. A miner simply leases a specified amount of hashing power without having a dedicated physical or virtual set up.

[2]The assumptions were borrowed from—https://bitcoinwisdom.com/bitcoin/difficulty

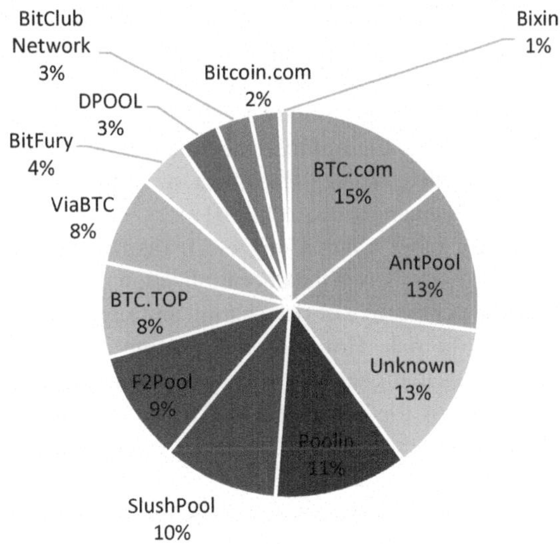

Fig. 1 Hash rate distribution (Data Source: https://www.blockchain.com/en/pools)

- Mining Pools—Mining pools are groups formed by miners to work together and pool their computing power. The main motive behind forming mining pools is to pool all of the resources to generate a higher hash power. In this model multiple miners contribute to generate the hash value and form a block; the block reward is split according to the miners' respective contributions in terms of computing power. The advantage of being a part of a mining pool is that it increases the probability of mining a block much faster due to the pooled computational power. However, one major drawback of joining a mining pool is the fee charged by the operator. Although the income earned by miners in the pool is steady, it is often quite less as an additional fee may be charged by the mining pool operator for operational expenses incurred. The below-depicted figure represents the hash rate distribution for Bitcoin mining among the existing mining pools as of May 2019.[3] BTC.com seems to be the most popular choice for miners and represents 15% of the total hash rate, whereas Bixin seems to be somewhat unpopular and represents only 1% of the total hash rate contribution. The Unknown contribution here represents the share contributed by un-identified/unknown mining pools. The difference in the hash rate contribution of different mining pools may be attributed to the fact that different pools generally have different characteristics such as different payout policies, different fee and so on. Before choosing a mining pool, one should take into consideration the factors that differentiate one pool from another (Salimitari et al. 2017) (Fig. 1).

However, mining pools present a somewhat contradictory concept to the whole idea of decentralization. Mining pools are a sort of centralization of computing power and are operated by a pool operator; this concept contradicts the original

[3]The data to create the chart was borrowed from—https://www.blockchain.com/en/pools

design of the Bitcoin system as a decentralized process. Miners participating in mining pools contribute their respective shares of computational power and these shares quantify the performance of individual miners in solving the block. Whenever one miner in the pool finds a valid block, the revenue is distributed among participants based on their individual contribution. Additionally, centralization in term of pools and geographical locations of mining operations impose control risks on miners (Hileman and Rauchs 2017).

On the other hand, Cong et al. suggest that the mining pool model does not necessarily challenge the decentralization concept of blockchains. They look at this from the perspective of risk sharing and suggest mining pools offer a way for miners to share the risk. However, risk sharing increases the competition among miners which leads to higher energy consumption. Empirical evidence lends support to their theory about the domination of decentralized blockchain systems over time by revealing the potential of miners for cross-pool diversification and fees charged by pools (Cong et al. 2019).

One of the approaches as a model for solving mining puzzles, i.e., solving blocks is to avoid the formation of mining pools. A large number of miners participating in a mining pool creates a target for attackers, and by doing so, a high volume of information and assets can be compromised. Furthermore, mining pools with centralized management could misuse the power to attack the network. Besides, mining pool administrators have the power to hide information, enforce transaction fees and lock-in periods. Additionally, if the miner finds a valid block, the miner may not necessarily send it to the mining pool. The miner may discard it due to some reasons; hence, this attack will decrease the efficiency of the mining pool because there will be no revenue based on discarded valid block. This is called the sabotage attack. The miner who finds a valid block but discards it will be rewarded due to the shares already submitted for the operator (Narayanan et al. 2016).

An alternative approach is to make sabotage attack profitable for miners which seems odd and interesting. In the current setting of mining pools, only the pool manager is in charge of rewards collection and requires all the participants to use the specific public key in their transactions. Also, the manager is the only one who has access to the private key and determines the distribution of newly minted coins. If the newly designed puzzles require knowledge of private key for the solutions, all the participants should have access to the private key. From this perspective, the model of a mining pool does not make much sense, and it would be profitable for the miner to continue mining individually and not in the pool (Narayanan et al. 2016).

2.2 *Hardware for Mining*

Now that we have introduced the basic concept of mining, how it works and the different ways you can mine, the next step is to have a look at the basic hardware requirements for mining and its evolution over time. A particular type of computer system is required for mining cryptocurrency. This system is generally referred to as a mining rig. This computer system can be built and operated solely for the purpose of mining, or it could be a system that fills other needs such as high-tech gaming or for running complex algorithms (Narayanan et al. 2016). However, Mining nowadays is being done by specialized mining rigs whose sole purpose is to mine cryptocurrency. A mining rig can be set up by procuring the following—Motherboard.

The Motherboard can be thought of as the brain of any computer system. The motherboard forms the foundation of the computer system into which everything is built and connected. It houses many of the critical components of any computer system such as the central processing unit, memory, graphics cards and allows connectivity to the other peripherals such as the mouse and keyboard.

CPU (Central Processing Unit): First Generation Mining

The next thing to think about is the CPU. The CPU can be thought of as the central nervous system of any computer system. It handles all of the complex procedures from starting up the system, running any sort of program or algorithm and is also responsible for making sure everything works seamlessly. Every CPU has a core which is responsible for running the processes; one core can only handle one process at a time. The more cores a CPU has, the more efficient it is. Nowadays, many processors (CPU) have multiple cores, for example—Intel's 8th generation processor has four. An average personal computer system generally has 2–4 cores depending on the CPU it houses. In terms of mining, a computer system having 4 cores means, the system can only solve 4 problems at a time. This is assuming that the sole purpose of the system is to mine, and no other programs or processes are being run simultaneously. The average computing power of a CPU is <30 MH/s (30 million hashes per second). It is quite evident from the facts presented that using a conventional computer system is not the most efficient way of mining any cryptocurrency, in order to really benefit from the process of mining, setting up a specialized computer system capable of handling multiple processes is of critical importance.

GPU (Graphical Processing Unit): Second Generation Mining

Graphical Processing Units served the purpose of enhancing the computational power of a CPU. Although a GPU is specifically built to handle geometry and other shape-based tasks such as 3-D graphics and visual effects, adding a GPU to a mining rig increases the total number of cores and hence, the computational power. An average GPU has about 20–30 cores and is considerably more efficient for mining purposes than a mining rig based only on a CPU. The average computing power ranges from 200 MH/s to 1 GH/s. GPUs are designed to run simultaneous programs; they have many logic units that can be utilized for a number of simultaneous SHA-256 computations and, moreover, most of the GPUs can be overridden to run faster than they are originally designed for (Bedford Taylor 2013). However, the main disadvantage of using a rig with GPU's is its massive energy consumption. Furthermore, GPUs do not have the most optimal thermal dynamics; they tend to overheat and designing a GPU mining rig can be a daunting process. Due to the energy consumption and high cost of ownership of GPU based rigs, mining using GPUs is becoming less and less economical. In recent times, miners have started to develop custom mining rigs which can be customized to the needs of the miner.

FPGA (Field Programmable Gate Array): Third Generation Mining
Mining rigs equipped with FPGA's can be programmed on a real-time basis. If a network changes its algorithm, an FPGA can be programmed to change with it. A typical FPGA mining rig can replicate multiple SHA-256 hash functions. FPGAs can be thought of like Lego blocks, i.e., an FPGA mining rig allows us to build and reconfigure using the same piece(s), it can be used to build any digital circuit, running different algorithms and software. Although the main aim of a rig equipped with FPGA is also to converge on a unique hash value as done by using CPUs and GPUs, it is much faster and more flexible as compared to using standalone CPU and GPU processors. The average computing power of FPGAs is significantly faster with a rate of 200 MH/s to 25 GH/s. FPGAs represent a significant jump in computational power and flexibility; however, the high cost of ownership is still a drawback as it only represents a marginal improvement over GPU based rigs. Furthermore, FPGAs are less accessible, you cannot buy them at most general electronic stores, and fewer individuals have the pre-requisite knowledge to set up a mining rig based on FPGAs as it requires expertise in computer engineering.

ASIC (Application-Specific Integrated Circuit): Fourth Generation Mining
Cryptocurrency mining today is being dominated by mining rigs equipped with ASICs. With specific reference to Bitcoin mining, ASIC chips are designed and built for the sole purpose of mining. ASICs particularly specialize in Bitcoin mining as the demand for computation power concerning Bitcoin mining outweighs the demand for mining other altcoins. The most efficient mining rigs nowadays are based on custom-designed ASICs, a single ASIC chip can produce computation power of up to 16 TH/s (Trillion hashes per second). However, the main drawback of ASICs is that they are quite costly to procure due to their specialized and time-consuming fabrication (Bedford Taylor 2013).

2.3 *Software for Mining*

Now that we have discussed the evolution of the cryptocurrency mining hardware, it is essential to talk about the software requirements that are necessary to connect the miners, blockchain and the mining pool. Software is necessary for transmitting information over the blockchain, to the miners and sending and receiving communications to and from the miners (Bhaskar and Chuen 2015). Mining software can operate on several different operating systems such as Windows, Mac OS, and Linux. The software supports the monitoring of vital mining statistics such as the temperature level of the hardware, hash rate, difficulty level and average speed of mining. Some of the most commonly used software for mining are

CGminer
CGminer has been quite a dominant force in the mining market as it has been continually evolving with the hardware. It is a modular opensource software written in C language that supports FPGA and AISC mining[4]; it is a cross-platform software which can be used on Windows, Mac OS and Linux. The main features of the software include monitoring hardware fan speed, temperature level, and remote interface capabilities.

BFGMiner
Another commonly used software is BFGMiner. Unlike CGminer which provided support for GPU mining prior to 2013, the BFGMiner is designed explicitly for FPGA and ASIC mining. Similar to CGminer, it is also written in the C language and is a cross-platform software. The main differentiating feature of this software is that it also includes a watchdog thread to detect computational malfunctions and restart the computation to spare the wastage of resources.

EasyMiner
This is the most user-friendly mining software available in the market. Unlike, CGminer and BFGMiner it is a graphic user interface (GUI) based mining software which is easy to use and understand. It is a perfect solution for miners who would prefer not to work with command-line based interface software such as the BFGMiner and CGminer. The features are similar to that of the previous two softwares (does not include watchdog thread); however, the EasyMiner is only available for the Windows operating system.

MultiMiner
Another user-friendly mining software with a GUI is the MultiMiner. MultiMiner is a simple yet powerful mining software which has some additional features when compared with the above software. This cross-platform software enables the miner to choose the mining strategy, it enables the miner to choose the currency to mine according to the hardware setup of the mining rig, and it also allows for advanced options for the experienced user such as direct access to the application programming interface (API).

3 Evolution of Mining

We have so far discussed the various hardware and software requirements to set up a well-equipped mining rig. However, mining in today's world has shifted from individuals towards professional mining farms and centers. Apart from the hardware and software considerations, setting up a well-equipped mining center needs to be a well thought out process as factors such cost of electricity, climate and network

[4]Versions later than 3.7.2, i.e., v3.8.0 and forthcoming updates (launched—November 2013) do not support GPU mining.

speed play a vital role in determining the efficiency of a mining operation. We touched upon the topic of cooling when talking about GPU mining; however, cooling plays a major role in mining with FPGAs and ASICs as well. More often than not overheating is a major cause for mining operations to stop and restart thereby decreasing the efficiency of the whole operation. Cooling is a major factor in mining operations; ideally a mining operation should be established in a cold climate where the cost of cooling would be low; however, in a situation where this is not possible, the operation could be established in a location where the cost of electricity is cheap. Furthermore, a fast and stable network connection is also vital as miners need to keep on top of the new blocks that are being formed as soon as they are announced to the network (Bhaskar and Chuen 2015).

If we look at the evolution of mining, it can be compared to the mining of precious metals such as gold and diamond. There are a number of similarities between mining cryptocurrency and other precious metals such as gold and diamond. The mining of precious metals in its infancy was done by individuals who started to use shallow pans to sort out the precious materials from gravel and other materials, an increase in demand led to more innovative ways of mining such as panning, by-product mining and ore processing (Narayanan et al. 2016). Similarly, cryptocurrency mining in its infancy was done by miners utilizing CPUs; however, as the difficulty level increased and CPU mining started to become non-viable, it led to the evolution of the methods, i.e., GPUs, FPGAs, and ASICs took over the cryptocurrency mining market. Additionally, mining moved from individuals to mining pools where computational power is pooled together to increase efficiency, and every miner is rewarded in accordance with the computing power contributed to solve the block. Furthermore, one striking similarity is how gold and diamond saw a rush in the nineteenth and twentieth century where individuals and organizations poured a lot of time, money and effort to look for prospects which would have an abundant store of these metals (Narayanan et al. 2016). Similarly, cryptocurrency mining has been undergoing a similar rush where the young and tech-savvy individuals are investing a lot of time, money and effort in order to find prospects of finding solutions to the block for the reward. However, like the precious metals, it is not necessary for every prospect to payout, as discussed miners may take several weeks and months to find a block solution. Moreover, there is much competition between miners and different mining pools to converge on the solution and claim the reward.

4 Factors Influencing Mining

The intensive hash calculations of the Bitcoin network reached approximately 26 quintillion hashes per second which require powerful resources of electricity equivalent to some countries. As a matter of fact, according to Bitcoin Energy

Consumption Index,[5] as of November 2017, the estimated annual electricity consumption of the Bitcoin exceeds more than 159 countries in the world. Cryptocurrency mining is not free from risks, and much like any other form of innovation, mining has some key risks factors that need a considerable amount of consideration. Although, the factors presented below are not exhaustive; however, they can be considered as the key factors affecting the economics of mining.

4.1 Energy Consumption and the Wastage of Resources

Digital wealth creation consumes a massive amount of energy and leaves a considerable environmental footprint. The current market value of cryptocurrencies accounts for more than 180 billion US dollars in which Bitcoin holds more than a 50% market share. Mining one US dollar worth of the Bitcoin consumes three times more energy than mining one US dollar worth of gold. The cryptocurrency industry currently utilizes more energy than some countries such as Denmark. According to a study conducted, mining operations of four commonly mined cryptocurrencies—Bitcoin, Ethereum, Litecoin, and Monero were responsible for 3–15 million tonnes of CO_2 emission during the years 2016–2018 (Krause and Tolaymat 2018). A similar study by Li et al. classified 13 types of cryptocurrencies based on their algorithms to estimate the electricity consumptions. Although the mining algorithms were quite different across selected coins, the consensus mechanisms were mainly Proof-of-Work and Proof-of-Activity (PoW/PoS-hybrid). Monero was selected as the cryptocurrency of interest to compute electricity consumption. The measurements to estimate energy consumption highlighted two features. First, each hashing algorithm has corresponding mining efficiency regardless of the coin. The second characteristic of the measurement is that the efficiency fluctuates across devices for the same hashing algorithm. This study demonstrated that hash rate and thermal design power (TDP) have a linear relationship and indicated that power is the determining factor in calculating the hash rate. In 2018, mining of Monero consumed around 700 GWh of electricity around the world. Mining operations in China consumed 33 GWh of the 700 GWh and consequently emitted 21,000 tons of CO_2 emission from April 2018 till the end of 2018 (Li et al. 2019).

The Bitcoin network includes more than 10,000 nodes, and each node represents a single piece of hardware or a connected hardware system (Mining rig). The precise estimation of the total electricity consumed by these nodes is still a matter of debate. The method to estimate total consumption is based on the efficiency of different machines in the network. Efficiency (J/GH) is defined in terms of joule per Giga hash. Calculation of lower bound electricity consumption is based on total network hash production and lower bound of efficiencies which is 0.098 J/GH. Based on this, the lower bound accounts for about 2.55 gigawatt (De Vries 2018). One of the major

[5]The index is available at—https://powercompare.co.uk/Bitcoin/

factors in the calculation of total electricity requirements is the cooling cost of the mining operation. A majority of cryptocurrencies are now mined in mining farms and centers with a massive number of connected devices which generate considerable heat. The cost of cooling based on local climate and cooling system must be taken into consideration when computing the total cost of energy. The power required to keep the mining centers cool and functional is estimated to impose an additional 30–50% increase in power consumption worldwide (Kampl 2014). The precise efficiency of mining facilities is also a grey area. One of the largest facilities to mine the cryptocurrencies is located in Ordos, China named—"Bitmain". Based on the different and conflicting reports of this center, the electric power consumption resides somewhere between 33 MW and 40 MW with more than 20,000 interconnected mining rigs (De Vries 2018).

Hayes proposes an economic-based approach to calculate electricity consumption in which cryptocurrencies are considered as commodities which are traded and produced in a competitive market. Based on market assumptions, miners continue the process until marginal costs of hash calculations equal their marginal product of coins (mined Bitcoins per day multiplied by US dollar exchange rate). This method is quite popular and is utilized as the base of the famous Bitcoin Energy Consumption Index (Hayes 2017).

As discussed earlier, mining is quite an expensive and tedious process; it involves the use of massive computational power. There is some discussion around the topic that cryptocurrency mining is a waste of energy as the energy spent on computing the SHA-256 algorithm is not useful for any other purpose. However, any type of payment network requires energy and electricity to function. Traditional payment systems such as ATMs, the printing of physical money as well as transporting this physical money requires a considerable amount of energy. The process of mining can be thought of as turning electricity into cash, when we mine cryptocurrency using hardware such as ASICs; we are basically using the computation power which consumes electricity as a way to turn it into cyber currency that can be sold for a certain amount of cash. Vranken in 2017 conducted a comprehensive literature review addressing the sustainability challenges of Bitcoin and development of hardware. The sustainability has been assessed from different perspectives—financial, environmental, ethical and economical. From the perspective of energy consumption and environmental impact, the primary estimations of electric power usage lie between 100 and 500 MW. When comparing the sustainability of mining operations in comparison with other gold mining and banking operations, the energy consumed per year for gold exploration and production, printing banknotes and minting coins, and banking operations in branches and ATMs are 500PJ (Petajoules), 40PJ and 2340PJ respectively which all are significantly higher than Bitcoin mining operation at 3-16PJ.

Based on the facts presented above, Bitcoin sustainability based on its energy consumption does not seem to be much of a concern. Some alternatives for the Proof-of-Work process with specific reference to the topic of energy consumption have also been proposed—Proof-of-Stake, Proof-of-Space are all alternatives to the POW process. Although these alternatives may consume less energy than the original POW process, there is still some concern regarding their security protocols. POW requires intensive hash calculations and consequently consumes considerable energy; however, the main advantage is that it also provides security in term of addressing double-spending concerns (McCook 2014; Vranken 2017).

As the number of miners participating in the network increases over time along with the level of mining difficulty, more powerful hardware systems including customized and optimized systems have been adopted by miners to increase the likelihood of finding valid solutions. If the cost of energy to produce cryptocurrencies is higher than the market value of mined coins, there would be no incentive to continue the process. O'Dwyer et al. compared the cost of energy for generating a Bitcoin by various advanced hardware systems including Core i7 950 (CPU), ATI 5770 (GPU), Digilent Nexys 2500 K (FPGA) and Monarch BPU 600 C (ASIC). When comparing the cost of energy with the exchange rate of a single unit of Bitcoin, the cost of mining on a CPU based mining hardware such as the Core i7 CPU was consistently higher than the US dollar worth of mined Bitcoin. In contrast, GPU hardware seemed to be somewhat profitable until mid-2013; however, post-2013, the cost of energy surpassed the value of the Bitcoin (Malone and O'Dwyer 2014).

Since the mining process requires intensive mathematical calculations to validate the transactions and generate cryptocurrency, we expect that the efficiency of mining will increase due to the advancements in hardware. However, the development and evolution of powerful hardware are taking place simultaneously with the augmentation in mining difficulty and the introduction of novel and complex algorithms which consume a massive amount of energy. There is no doubt that cryptocurrency has certain negative impacts on the environment as securing energy resources for the digital industry mostly depends on the availability of fossil fuels which questions the sustainability of technological process such as mining. One of the challenges of the cryptocurrency mining industry is that despite of consuming a massive amount of resources, it is unable to expand local economies in terms of contributing in the form of taxes or jobs by its operations. In March 2018, Plattsburgh; a small city in the US (near the Canadian border) enacted a moratorium in which expanding virtual currency operations have been banned due to the prospect of cryptocurrency miners consuming large amounts of electricity. Since the power consumption was about to exceed its threshold, this ban was to protect the local population from future pressure of higher tax in the purpose of providing more power generation facilities (D'Ambrosio 2018). After 1 year, the limitation on virtual currency operation has been lifted in the city, but still, the officials stated that cryptocurrency miners are responsible for the cost of excessive power usage.

As mentioned earlier, the profitability of cryptocurrencies depends on the cost of mining. Clean energy as an alternative source of energy could guarantee the long-term sustainability of value production of cryptocurrency mining operations. The first clean-energy operated mining hub was established in the Japanese city of Kazuno. The city of Kazuno provides natural resources for the mining operation. Another strategy adopted by Swiss startup Envion is to manufacture a transportable mining system to be able to use the cheapest clean energy provided anywhere in the

Profit/Loss	Mined $ Value	Mined BTC Value	Electricity Costs $ Value
-$2.69/Day	$1.58/Day	0.0003 BTC/Day	$4.26/Day
-$79.95/Month	$47.93/Month	0.0087 BTC/Month	$127.87/Month
-$959.34/Year	$575.12/Year	0.1043 BTC/Year	$1534.46/Year

Fig. 2 Energy consumption (Data source: https://www.buybitcoinworldwide.com/mining/hardware/)

world (Girard 2018). However, this strategy is not the most ideal due to the various legal restriction in place. As a matter of fact, Envion is now facing various regulatory accusations.

Now to discuss profitability and energy consumption in terms of numbers, let's take a look at an example—The most advanced ASICs nowadays claim to have a hash rate of about 16 TH/s and an average energy consumption of 1480 watts, the current price of Bitcoin is quoted as 5516.02 (As of 2nd May 2019) and a 0.4527678% daily increase in network hash rate. If we assume the cost of electricity, i.e., KW/h in USD to be 0.12. Then using the above assumptions, we can compute the daily, monthly and yearly profit/loss (Fig. 2).[6]

As you can see using an ASIC with the above specification is not really profitable given the current price of Bitcoin and electricity. We can observe from the table that the daily project loss is $2.69, the monthly projected loss is $79.95, and the yearly projected loss is $959.34. This leads us to another risk factor to consider—volatility of the cryptocurrency being mined. Price volatility is a vital risk factor to consider as a change in the price of the cryptocurrency can have a massive impact on any mining operation.

4.2 Volatility

There are several concerns regarding the risks of Bitcoin mining, due to the arguments concerning environmental impacts of mining operation in terms of energy consumption and CO_2 footprint, there is a risk of restrictions on mining operations by governments and local authorities. For example, miners may be held responsible for a surge in energy prices. Furthermore, governments may increase the taxation of mining revenues as the mining industry is becoming a well-established industry with more and more miners pouring in from all over the world. From the operational perspective, small and individual miners are more concerned about operational risk factors than large miners; however, the volatility of the cryptocurrency of interest is a vital risk factor to consider for large and small miners alike. Unexpected electricity price hike in the future, cyber-attacks and aggressive competition among miners are all valid risk factors; however, the volatility, profitability and decreasing block

[6]Computation was done using the calculator available at https://www.buybitcoinworldwide.com/mining/hardware/

rewards have most miners on edge as cryptocurrency mining operations may not be economical going forward (Hileman and Rauchs 2017).

An interesting perspective is presented in Liu and Tsyvinski's (2018) research, they utilize and compare three coins and their trade-offs with traditional financial markets including stocks, currencies, and precious metals. They suggest that the risk and return relationship of cryptocurrencies is entirely diverse from the capital markets. Furthermore, macroeconomic risk factors and volatility in the stock market are not determinants of returns in the cryptocurrency market. The cryptocurrency included in the research is not affected by the returns of other financial markets including currencies and metal commodities. The volatility in the cryptocurrency market is determined by factors which are specific to the cryptocurrency market, and one of the main factors contributing to volatility is the time-series momentum effect which relates to investor attention and behaviour (Liu and Tsyvinski 2018).

4.3 Mining Puzzles/Solving Blocks

Another factor affecting mining is the difficulty level of mining puzzles. As discussed, the main aim of miners is to try and come up with the most efficient way to solve the blocks for the block reward. Since all the nodes in the network must verify the validity of the puzzle, mining puzzles should be quickly verifiable, and the difficulty of mining should be adjusted based on the number of members in the network due to higher contributed hash power. Modification of puzzle difficulty protects the process against attacks on the block-chain and simultaneously attempts to maintain an average rate of solving the block. Furthermore, the proportionality of the hash power and the probability of solving the block are correlated, i.e., the more powerful the hardware, the more chances of success to solve the puzzle. This presents an incentive to all miners to participate in the network, and ensures every miner has a chance of winning the next puzzle solution based on their contributed hash powers and the probability of solving is independent of how much work miners have spent on the solving (memory-less and progress free) (Narayanan et al. 2016). The SHA-256-algorithm satisfies the mentioned requirements (Mukhopadhyay et al. 2016).

As mentioned before, cryptocurrency mining was initially done by conventional computer hardware such as CPUs and GPUs; however, it rapidly evolved to more sophisticated and efficient hardware such as ASIC chips. The current difficulty of solving the block does not make conventional mining economical anymore. This eventually may be detrimental to the democratic settings of the Bitcoin system due to the fact that challenges and cost of mining process eliminate small miners and the limited number of powerful miners controls the whole system. So the question arises here is the possibility of designing alternative puzzles which could still be economically solvable by individual users who use general-purpose computers (ASIC resistant) in comparison with big miners with highly customized and specialized hardware (Narayanan et al. 2016).

The most popular puzzles to beat the ASIC optimized hardware are memory-hard puzzles which require a large amount of memory instead of huge CPU time. Scrypt is a popular memory-hard puzzle which is mostly used in Litecoin. The Scrypt algorithm came before developing the Bitcoin system and was used in password hashing. Scrypt is utilized in password hashing by developing a secure system which increases the time for attackers with customized hardware to breach the password. Furthermore, the second approach to design ASIC resistant puzzles is X11 which is not very well developed in comparison with Scrypt. It is a combination of 11 hash functions designed to increase the complexity of using efficient ASIC. Another approach to make optimized mining hardware less efficient is designing moving target puzzles. These puzzles change periodically and make the ASIC hardware ill fitted to deal with the previous puzzle as they are constantly changing (Narayanan et al. 2016).

References

Bedford Taylor M (2013) Bitcoin and the age of bespoke silicon. 2013 international conference on compilers, architecture and synthesis for embedded systems (CASES), pp 1–10. https://doi.org/10.1109/CASES.2013.6662520

Berentsen A, Schar F (2018) A short introduction to the world of cryptocurrencies. Review 100 (1):1–19. https://doi.org/10.20955/r.2018.1-16

Bhaskar ND, Chuen DLK (2015) Bitcoin mining technology. In: Handbook of digital currency, pp 45–65. https://doi.org/10.1016/B978-0-12-802117-0.00003-5

Cocco L, Marchesi M (2016) Modeling and simulation of the economics of mining in the bitcoin market. PLoS One 11(10):e0164603. https://doi.org/10.1371/journal.pone.0164603

Cong LW, He Z, Li J (2019) Decentralized mining in centralized pools (no. 0898–2937). National Bureau of Economic Research

D'Ambrosio D (2018) Plattsburgh turns back invasion of bitcoin miners. https://www.forbes.com/sites/danieldambrosio/2018/10/31/plattsburgh-turns-back-invasion-of-bitcoin-miners/#1ce3adb94b5b

De Vries A (2018) Bitcoin's growing energy problem. Joule 2(5):801–805

Girard L (2018) Environmental impacts of cryptocurrency mining

Hayes AS (2017) Cryptocurrency value formation: an empirical study leading to a cost of production model for valuing bitcoin. Telematics Inform 34(7):1308–1321

Hileman G, Rauchs M (2017) Global cryptocurrency benchmarking study. Cambridge Centre for Alternative Finance, p 33

https://bitcoinwisdom.com/bitcoin/difficulty (n.d.) BitcoinWisdom. Available via BitcoinWisdom website: https://bitcoinwisdom.com/bitcoin/difficulty

https://powercompare.co.uk/bitcoin/ (n.d.). https://powercompare.co.uk/bitcoin/. Accessed 01 Jan 2019

https://www.blockchain.com/en/pools (n.d.) Blockchain. Available via Blockchain website: https://www.blockchain.com/en/pools

https://www.buybitcoinworldwide.com/mining/hardware/ (n.d.) buybitcoinworldwide. Available from buybitcoinworldwide website: https://www.buybitcoinworldwide.com/mining/hardware/

Information systems security: 10th international conference, ICISS 2014, Hyderabad, India, December 16–20, 2014. Proceedings (1st edition) (2015) Springer, New York

Kampl A (2014) Analysis of large-scale Bitcoin mining operations. http://www.immersion-cooling.com/publications/Analysis_of_Large-Scale_Bitcoin_Mining_Operations.pdf

Krause MJ, Tolaymat T (2018) Quantification of energy and carbon costs for mining cryptocurrencies. Nat Sustain 1(11):711

Li J, Li N, Peng J, Cui H, Wu Z (2019) Energy consumption of cryptocurrency mining: a study of electricity consumption in mining cryptocurrencies. Energy 168:160–168

Liu Y, Tsyvinski A (2018) Risks and returns of cryptocurrency. National Bureau of Economic Research
Malone D, O'Dwyer KJ (2014) Bitcoin mining and its energy footprint. 25th IET Irish Signals & Systems Conference 2014 and 2014 China-Ireland International Conference on Information and Communities Technologies (ISSC 2014/CIICT 2014), pp 280–285. https://doi.org/10.1049/cp.2014.0699
McCook H (2014) An order-of-magnitude estimate of the relative sustainability of the Bitcoin network. https://www.academia.edu/7666373/an_order-of-magnitude_estimate_of_the_relative_sustainability_of_the_bitcoin_network_-_2nd_edition. Accessed 18 May 2018
Mukhopadhyay U, Skjellum A, Hambolu O, Oakley J, Yu L, Brooks R (2016) A brief survey of cryptocurrency systems. IEEE, pp 745–752
Nakamoto S (2008) Bitcoin: a peer-to-peer electronic cash system, p 9
Narayanan A, Bonneau J, Felten E, Miller A, Goldfeder S (2016) Bitcoin and cryptocurrency technologies: a comprehensive introduction. Princeton University Press, Princeton, NJ
Salimitari M, Chatterjee M, Yuksel M, Pasiliao E (2017) Profit maximization for bitcoin pool mining: a prospect theoretic approach. 2017 IEEE 3rd international conference on collaboration and internet computing (CIC), pp 267–274. https://doi.org/10.1109/CIC.2017.00043
Vranken H (2017) Sustainability of bitcoin and blockchains. Curr Opin Environ Sustain 28:1–9
Yli-Huumo J, Ko D, Choi S, Park S, Smolander K (2016) Where is current research on blockchain technology?—a systematic review. PLoS One 11(10):e0163477. https://doi.org/10.1371/journal.pone.0163477

Regulating Bitcoin: A Tax Case Study

Margaret Ryznar

Abstract This chapter adapts the Coffee bonding theory to the modern context of bitcoin, using tax as a case study. As the theory predicts, tax authorities may be able to increase the legitimacy of bitcoin by improving tax compliance and reducing tax evasion. Thus, while the Coffee theory arose two decades ago to explain the cross-listing of foreign company shares, it has implications for the modern context of bitcoin.

1 Introduction

Cross-listing of foreign company shares on American exchanges has been occurring for decades. Commentators have offered several explanations for this phenomenon, including the prominent Coffee bonding theory that there are legitimacy gains to adopting American securities laws through cross-listing (Coffee 2002). This theory about regulation has direct implications in the bitcoin context. Satoshi Nakamoto, a pseudonym for an unknown person, designed bitcoin and introduced it in a 2008 white paper titled "Bitcoin: A Peer-to-Peer Electronic Cash System" (Nakamoto 2008). At its essence, bitcoin is a cryptocurrency, which is a digital currency issued electronically by a computer program. Bitcoin has a predetermined cap of 21 million (Groshoff 2014). To implement bitcoin, Nakamoto devised the first blockchain to solve the double-spending problem for digital currency so that people cannot spend the same money twice (Shackelford and Myers 2017).

As bitcoin gained prevalence in the United States, it has drawn the attention of regulators. U.S. agencies considering bitcoin issues have included the Securities and Exchange Commission (SEC), the Commodity Futures Trading Commission (CFTC), the Financial Crimes Enforcement Network (FinCEN), and the Internal Revenue Service (IRS) (Burks 2017). These agencies have differed in their treatments of virtual currency, and comprehensive regulation has been unsuccessful to date (Burks 2017). Challenges include bitcoin's rapid growth and anonymity.

M. Ryznar (✉)
McKinney School of Law, Indiana University, Indianapolis, IN, USA
e-mail: mryznar@iu.edu

S. Goutte et al. (eds.), *Cryptofinance and Mechanisms of Exchange*, Contributions to Management Science, https://doi.org/10.1007/978-3-030-30738-7_4

Thus far, many bitcoin users have resisted any regulation (Zaytoun 2019). Yet, regulation may provide bitcoin with benefits that include legitimacy, as articulated by the Coffee theory that arose two decades ago to explain the cross-listing of shares in terms of regulatory benefits. This chapter explores the adaptability of this theory to bitcoin, examining the benefits of regulation in terms of increasing the legitimacy of bitcoin. In particular, this chapter uses the taxation of bitcoin in the United States as a case study, considering involuntary and voluntary noncompliance with the tax laws and their consequences for bitcoin's legitimacy.

2 The Coffee Bonding Theory

The Coffee bonding theory initially arose in the context of the cross-listing of foreign stocks (Coffee 2002). If they meet the regulatory requirements, foreign companies have the opportunity to cross-list their shares on a foreign stock exchange, such as the New York or London stock exchange. This allows foreign companies to list their shares on their own domestic exchange as well as a foreign exchange, usually in the United States (Pine 2010).

Cross-listing has traditionally been explained as an attempt to break down market segmentation and to increase investor recognition of the cross-listing firm (Baker et al. 2002). Another prominent explanation put forth by Professor Coffee is "bonding." Under this theory, issuers migrate to U.S. exchanges because by voluntarily subjecting themselves to higher disclosure standards and greater threat of enforcement both by public and private enforcers, they partially compensate for weak protection of minority investors under their own jurisdictions' laws and thereby achieve a higher market valuation (Coffee 2002). In other words, a firm's decision to cross-list on a U.S. exchange subjects it to a set of new disclosure and legal requirements (Ferris et al. 2009). Firms are thus choosing to "rent" the securities laws of other countries under the bonding theory (Pine 2010).

Although compliance with foreign regulators to cross-list is an expensive endeavor for companies (Saudagaran 1988), exposure to an international capital market can induce changes in corporate governance and improve investor perception of the quality of its governance (Ferris et al. 2009). A U.S. listing can also reduce the extent to which controlling shareholders can engage in expropriation and thereby increase the firm's ability to take advantage of growth opportunities (Doidge et al. 2004). In other words, U.S. regulation can provide increased legitimacy (Coffee 2002). This Coffee theory has implications for bitcoin.

From its beginning, bitcoin has suffered from legitimacy concerns stemming from the lack of regulation. Without a central bank underlying cryptocurrencies, many people find it difficult to trust bitcoin as a currency (Kearns 2013). The volatility of bitcoin's price also makes it hard to trust (Christopher 2016). Bitcoin is neither intrinsically valuable, like gold, nor is it rooted in a commodity expressing a certain purchasing power (Plassaras 2013). There might be some value resulting from its scarcity, but it is an artificial scarcity (Burge 2016). Generally, bitcoin as a currency is not regulated like stocks and futures, and lighter regulation facilitates price manipulation (Markham 1991). Bad actors can manipulate the price of cryptocurrencies and then cash out before other investors catch on. There are also concerns about initial coin offerings of bitcoin, with the main reason for going public being for insiders to cash out. Additional concerns arise regarding a bitcoin bubble (Giancarlo 2018). All of these contribute to bitcoin's price volatility (Ly 2014).

Tax compliance issues, some rising to the level of tax evasion, have also undermined the legitimacy of bitcoin, which has prompted efforts to increase regulation aimed at solving them. The next part of this chapter examines the benefits of tax regulation for bitcoin, which are consistent with the Coffee bonding theory.

3 A Tax Case Study

The anonymity of bitcoin can facilitate tax evasion, which has attracted illegitimate users among legitimate ones (Foley et al. 2019). Yet, virtual currencies are a potential source of highly secure, private, and fluid transactions. By providing better guidance that supports the legitimate purpose of virtual currencies, tax authorities such as the IRS can empower users to take advantage of the benefits that virtual currencies offer. Perhaps more importantly to the IRS, proper guidance could improve reporting of virtual currency gains, thereby increasing tax revenue.

Tax compliance can always be improved, but compliance issues particularly abound in the anonymous world of bitcoin, which is devoid of connections to governments and mortar banks. This has undermined the legitimacy of bitcoin (Gruber 2013).

Tax regulators have started to address the issues stemming from bitcoin's unique characteristics. Already, several countries including the United States have collaborated to increase enforcement (IRS 2018). The IRS may also soon be working to develop its own policies on virtual currencies (Information Reporting Advisory Committee 2018). Consistent with Coffee's theory, there would be added benefits to such regulation for bitcoin, including a legitimacy boost.

When it comes to the taxation of bitcoin, there are several ways to improve compliance with the tax laws to increase the legitimacy of bitcoin (Edward 2006). American authorities have already moved to implement some of them in an effort to improve tax noncompliance and reduce the use of bitcoin for tax evasion. It is important to address both involuntary and voluntary noncompliance given their differences.

3.1 Involuntary Noncompliance

Much involuntary noncompliance with the U.S. tax laws on bitcoin stems from confusion. The federal tax code is complex, yet it becomes even more so when applied to bitcoin. Even when people want to comply with the tax laws regarding bitcoin, they may have trouble doing so due to this complexity.

There are two main ways to acquire bitcoin—to buy it on an exchange such as Coinbase or to earn it by processing bitcoin transactions, called "mining" (Akins et al. 2014). Mining immediately triggers tax consequences, with the fair market value of the coins mined included in gross income. If the bitcoin is not liquidated at the time it was mined, then it becomes a capital asset and receives the same tax treatment as buying it on an exchange. While buying bitcoin has no tax consequences, selling it can yield capital gains or losses like other property investments.

This is the tax treatment outlined by IRS Notice 2014–2021 ("Notice"), the only guidance to date on the income taxation of virtual currency. In it, the IRS made clear that it treats virtual currency as property instead of currency. The Notice describes how existing tax principles apply to transactions using virtual currency and answers a variety of common questions relating to the income tax treatment of virtual currency gains or losses. However, it left many unanswered questions. Guidance from the IRS in addition to its 5-year-old Notice could improve reporting of virtual currency gains, thereby increasing tax revenues overall (U.S. Gov't Accountability Office 2013). The Treasury Inspector General for Tax Administration (TIGTA) has thus suggested in a recent report the need for better tax guidance (Treasury Inspector General for Tax Administration 2016).

Not only is it difficult for taxpayers to understand the tax law regarding bitcoin, but it may be difficult to comply given the nature of bitcoin, which includes a currency function. To simplify tax compliance, lawmakers in the future may choose to consider a de minimis exception for bitcoin transactions. Exempting gain on a transaction below a certain threshold would dispose of a huge segment of virtual currency transactions because smaller transactions would not be subject to taxation.

With such a de minimis exception, casual bitcoin users could therefore buy a certain amount of goods or services with virtual currency without any tax consequences, but the primary limitation would be the potential volatility of the value of bitcoin that might wildly fluctuate below and above the de minimis exception. Nonetheless, the threshold should be high enough to dispose of a large number of routine consumer transactions.

Consider, for example, the oft-envisioned future of bitcoin, where users pay for daily small purchases, such as a cup of coffee, directly from their virtual wallets. Indeed, coffee seller Starbucks plans to accept payment in bitcoin starting in 2020. Without a de minimis exception, purchasing a cup of coffee would be a taxable event, requiring taxpayers to calculate their gain or loss on the transaction. A de minimis exception would eliminate this result.

The unsuccessful Cryptocurrency Tax Fairness Act, proposed in the United States in September 2017, contained such a de minimis exception. Under its approach, any transaction resulting in $600 or less of gain would be excluded from taxation. The

United Kingdom has already adopted a de minimis exception, although its threshold is much higher—£11,700 of gain in cryptocurrency transactions is tax-free under certain conditions (Crypto Daily 2018).

Overall, taxpayers would benefit from the simplified process resulting from a de minimis exception and the IRS would still capture significant revenue from large virtual currency transactions. Such decreased tax regulation in low-value transactions would raise the efficiency of using bitcoin without jeopardizing its legitimacy. Meanwhile, high-volume bitcoin users would benefit from additional information beyond Notice 2014–2021 to assist with tax compliance. These changes would provide a boost to the legitimacy of bitcoin by increasing tax compliance.

3.2 Voluntary Noncompliance

Voluntary noncompliance with the tax laws may rise to the level of tax evasion, which is a felony crime in the United States punishable by a $100,000 fine and 5 years imprisonment per 26 U.S. Code §7201. Thus, more so than involuntary noncompliance stemming from confusion, bitcoin's use to evade taxes undermines its legitimacy.

While there is an incentive to report bitcoin losses to claim tax deductions, the same is not true of bitcoin gains. As a result, some bitcoin users intentionally do not report their gains. Despite the existence of penalties for underreporting tax liability in the United States, they are difficult to apply due to the anonymity of bitcoin. This has led to bitcoin's ability to function like Swiss banks (Morris 2014). Voluntary noncompliance with the tax laws costs the U.S. Treasury billions of dollars each year (Marian 2013).

To combat the anonymity surrounding bitcoin, the IRS has made progress in establishing its authority to summon records from a virtual currency platform through a "John Doe" anonymous summons. In U.S. v. Coinbase, Inc., the IRS served a summons on Coinbase seeking information on essentially all of its users. The IRS ultimately limited its request to information for users with the equivalent of $20,000 in one transaction—around 10,000 users. The district court enforced, in part, the narrowed summons, ordering Coinbase to produce records revealing the name, taxpayer identification number, birth date, address, transactions logs, and account statements of certain users. While the scope of the summons was significantly narrowed, it still represented a victory for the IRS. The ability of the IRS to gather records necessary to examining a taxpayer's virtual currency transactions will only increase the frequency and accuracy of reporting of gains and losses.

As in other tax contexts, third party reporting could also improve compliance (Hatfield 2015). Coinbase, the largest bitcoin platform, currently issues voluntary Form 1099-K to a select group of users—those with at least 200 annual transactions totaling at least $20,000 who use Coinbase for business purposes (Coinbase, 1099-K Tax Forms 2018). In order to provide the IRS with a better picture of the true scope of bitcoin transactions, all virtual currency platforms can be required to report user activity on more than just large-volume business users.

In sum, noncompliance with tax laws regarding bitcoin can be voluntary or involuntary. Voluntary noncompliance, in particular, has given bitcoin the reputation of facilitating tax evasion, undermining the legitimacy of bitcoin. Despite the novelty surrounding bitcoin and other virtual currencies, traditional tax compliance methods can be adopted to address many of these noncompliance issues. Such tax regulation would increase bitcoin's legitimacy, as the Coffee bonding theory would predict.

4 Conclusion

While the Coffee bonding theory originally arose two decades ago in the context of the cross-listing of foreign stocks, it also has implications today for bitcoin. In the same way as it does for foreign stocks, regulation legitimizes bitcoin to a certain extent, particularly important given its start as an anonymous cryptocurrency for illegal activities (Foley et al. 2019). Historically, the legitimizing effect of regulation has brought some value (Coffee 2002). On the other hand, criticism has generally targeted the enforcement of any regulation. For example, there is the possibility of bias in enforcement of the laws (Heminway 2003). Furthermore, there are separate critiques regarding over-regulation of the business environment (Woody 2012). These concerns regarding regulation no doubt hold true in the bitcoin context, but must be considered alongside the benefits of regulation, including those predicted by the Coffee bonding theory.

References

Akins BW, Chapman JL, Gordon JM (2014) A whole new world: income tax considerations of the bitcoin economy. Pittsbg Tax Rev 12:25

Baker HK, Nofsinger JR, Weaver DG (2002) International cross-listing and visibility. J Financ Quant Anal 37:495

Burge ME (2016) Apple pay, bitcoin, and consumers: the ABCs of future public payments law. Hastings Law J 67:1493

Burks C (2017) Bitcoin: breaking bad or breaking barriers? N C J Law Technol 18:244

Christopher CM (2016) The bridging model: exploring the roles of trust and enforcement in banking, bitcoin, and the blockchain. Nev Law J 17:139

Coffee JC (2002) Racing towards the top?: the impact of cross-listings and stock market competition on international corporate governance. Columbia Law Rev 102:1757

Coinbase, 1099-K Tax Forms (2018) https://support.coinbase.com/customer/en/portal/articles/2721660-1099-k-tax-forms. Accessed 30 Dec 2018

Crypto Daily (2018) Crypto tax and ICO regulations in the United Kingdom. https://cryptodaily.co.uk/2018/08/crypto-tax-and-ico-regulations-in-the-united-kingdom

Doidge C, Karolyib GA, Stulz RM (2004) Why are foreign firms listed in the U.S. worth more? J Financ Econ 71:205

Edward CK (2006) Structural laws and the puzzle of regulating behavior. Northwest Univ Law Rev 100:655

Ferris SP, Kim KA, Noronha G (2009) The effect of crosslisting on corporate governance: a review of the international evidence. Corp Gov 17:338

Foley S, Karlsen JR, Putniņš TJ (2019) Sex, drugs, and bitcoin: how much illegal activity is financed through cryptocurrencies? Rev Financ Stud 32:1798

Giancarlo JC, written testimony before the U.S. Senate Agriculture, Nutrition, and Forestry Committee, Washington, D.C. (15 Feb 2018). http://www.cftc.gov/PressRoom/PressReleases/opagiancarlo38

Groshoff D (2014) Kickstarter my heart: extraordinary popular delusions and the madness of crowdfunding constraints and bitcoin bubbles. William and Mary Bus Law Rev 5:489

Gruber S (2013) Note, trust, identity, and disclosure: are bitcoin exchanges the next virtual havens for money laundering and tax evasion? Quinnipiac Law Rev 32:135

Hatfield M (2015) Taxation and surveillance: an agenda. Yale J Law Technol 17:319

Heminway JM (2003) Save Martha Stewart? Observations about equal justice in U.S. insider trading regulation. Texas J Women Law 12:247

Information Reporting Advisory Committee October 2018 Public Report 73 (2018). https://www.irs.gov/pub/irs-pdf/p5315.pdf

IRS, Joint Chiefs of Global Tax Enforcement (2018). https://www.irs.gov/compliance/joint-chiefs-of-global-tax-enforcement

Kearns J (2013, December 4) Greenspan says bitcoin a bubble without intrinsic currency value, Bloomberg. www.bloomberg.com/news/2013-12-04/greenspan-says-bitcoin-a-bubble-without-intrinsic-currency-value.html

Ly MK (2014) Note, coining bitcoin's "legal-bits": examining the regulatory framework for bitcoin and virtual currencies. Harv J Law Technol 27:587

Marian O (2013) Are cryptocurrencies super tax havens? Mich Law Rev First Impressions 112:38

Markham JW (1991) Manipulation of commodity futures prices—the unprosecutable crime. Yale J Regul 8:281

Morris P (2014) Bitcoin: lifting the veil. Westlaw J Bank Lender Liabil 19(25):1

Nakamoto S (2008) Bitcoin: a peer-to-peer electronic cash system. https://bitcoin.org/bitcoin.pdf

Pine KW (2010) Lowering the cost of rent: how IFRS and the convergence of corporate governance standards can help foreign issuers raise capital in the United States and Abroad. Northwest J Int Law Bus 30:483

Plassaras NA (2013) Comment, regulating digital currencies: bringing bitcoin within the reach of the IMF. Chic J Int Law 14:377

Saudagaran SM (1988) An empirical study of selected factors influencing the decision to list on foreign stock exchanges. J Int Bus Stud 19:101

Shackelford SJ, Myers S (2017) Block-by-block: leveraging the power of blockchain technology to build trust and promote cyber peace. Yale J Law Technol 19:334

Treasury Inspector General for Tax Administration (2016) As the use of virtual currencies in taxable transactions becomes more common, additional actions are needed to ensure taxpayer compliance. https://www.treasury.gov/tigta/auditreports/2016reports/201630083fr.pdf

U.S. Gov't Accountability Office (2013) Virtual economies and currencies: additional IRS guidance could reduce tax compliance risks, pp 3–5. http://www.gao.gov/assets/660/654620.pdf

Woody K (2012) Conflict minerals legislation: the SEC's new role as diplomatic and humanitarian watchdog, 81 Fordham L. Rev. 1315

Zaytoun HS (2019) Cyber pickpockets: blockchain, cryptocurrency, and the law of theft. N C Law Rev 97:395

Are Cryptocurrencies Truly Trustless?

Usman W. Chohan

Abstract A common narrative of cryptocurrencies presents them as "trustless," decentralized, and autonomous systems. The "trustlessness" is meant to suggest lack of need for third-party verification in blockchain technologies, but the term has been somewhat conflated with broader connotations of "trust." This chapter draws out that nuance, and critically assesses that claim, by emphasizing the human element of trust in cryptocurrencies across various contexts. It does so by highlighting four activities that require both direct human intervention and direct human participation, including: "hard forks" to directly change protocols, the management of cryptocurrency exchanges, the emission of ICOs, and investor recourse to traditional governance institutions including courts of law. The findings of the chapter therefore suggest that the cryptocurrency space is not "trustless" in every sense as it is still reliant on the trust-element in human agency and structure.

1 Introduction

As the appeal of cryptocurrencies has grown, there are certain adjectives that have been ascribed to them to emphasize their distinctions with the monetary systems of traditional finance. Four of those adjectives are decentralized, autonomous, immutable, and trustless. Decentralization is meant to refer to the distribution of information across a panoply of users on the blockchain system without centralized control (see Chohan 2017a). Autonomous refers to the ability of the system to work under pre-programmed algorithms without the need for constant human interventions or oversight (Chohan forthcoming-b). Both of these terms are encapsulated in the Decentralized Autonomous Organization (DAO), a short-lived experiment based on an offshoot of blockchain technology which, despite practical failings (hacks), constituted a conceptually rich topic of enquiry (see Chohan 2017b). Immutability refers to the fact that tampering and erasure cannot generally occur once information is put into the blocks of a chain.

U. W. Chohan (✉)
UNSW School of Business, The University of New South Wales, Sydney, NSW, Australia
e-mail: u.chohan@adfa.edu.au

S. Goutte et al. (eds.), *Cryptofinance and Mechanisms of Exchange*, Contributions to Management Science, https://doi.org/10.1007/978-3-030-30738-7_5

Yet the fourth adjective, and the one of greatest import to this chapter, is trustlessness, It does not refer to the absence of trust, but rather to the absence of a need for trust in the system. Specifically, it was intended to refer to the lack of third-party verification requirements that blockchain allows for when it solves the double-spending problem (Chohan 2017c). However, the term has been somewhat conflated with broader connotations of "trust." By extension, an insinuation in the popular discourse (see discussions in Chohan 2019a, b) has been that blockchain technologies have no significant trust-requirements of any sort.

In other words, a point of nuance has been lost. Although third-party verification is effectively not required in the blockchain architecture and it is therefore "trustless" in one sense, there is indeed a significant element of trust that permeates the socioeconomic realm of cryptocurrency transactions. Whenever trust is compromised in the commercial transactions of cryptocurrencies, the risk to their viability as monetary instruments grows. Infractions along those lines might include: malpractice in cryptocurrency exchanges (referred to henceforth in portmanteau as "cryptoexchanges"), hacking, thefts, deception, money laundering, Ponzi schemes, and fraud (see Chohan 2018a, b, c). As such, there is a need to correct the dissemination of inaccurate representations of cryptocurrencies among a public audience, which includes investors who have made decisions to buy and sell cryptocurrencies without being entirely aware of the nature of "trustlessness."

Why should so much attention be paid to what might appear to be a semantic issue in non-specialist popular discussions? To put it plainly, trust is the bedrock of all sustained commercial life. Cryptocurrencies have suffered a dent in their popular appeal and credibility because of issues of trust which, although not emanating directly from the structure of blockchains, have made themselves manifest in the social realm where anonymous users and groups have compromised trust to the ultimate detriment of investors and network participants (Chohan 2018a, b, c). By extension, this has led to reputational damage to cryptocurrencies to a degree that might threaten their long-term viability, since their wider adoption is necessary for them to become effective complements (and for idealists: substitutes) to traditional monetary instruments.

With this in mind, the aim of this chapter is to critically assess and draw distinctions in the concept of "trustlessness" as it is understood in the realm of cryptocurrencies (see Chohan 2019a). To do so, the chapter first reviews the academic literature as it has referred to the trustless element in blockchain, including through normative expectations about the advantages that it is thought or expected to bring. The chapter then analyses and highlights four activities that require both direct human intervention and participation, including: "hard forks" to directly change protocols, the management of cryptocurrency exchanges, the emission of initial coin offerings (ICOs), and investor recourse to traditional governance institutions including courts of law. These activities emphasize the fact that, the lack of need for third-party verification notwithstanding, the cryptocurrency space is not truly "trustless" as it is still reliant on the trust-element in human agency and within human trust-bound structures. A final section concludes with a discussion of the cryptoanarchist philosophical implications of trust in trustless systems.

2 The Promise de Trustlessness

It is of great appeal to a segment of cryptocurrency participants that the blockchain system operates in a "trustless" manner without the need for third-party verification. Early academic work on blockchain technology has generally commended this trait, and highlighted it as a significant advantage of cryptocurrencies vis-à-vis the traditional financial system. For example, Forogolu and Tsilidou emphatically state the point that "the whole thing about blockchain-based architectures is that they allow trustless transactional activity," (Foroglou and Tsilidou 2015, p. 2), and Kiviat makes the bold assertion that "trustless means that—for the first time in history—exchanges for value over a computer network can be verified, monitored, and enforced without the presence of a trusted third party or central institution," (Kiviat 2015, p. 574). Bahga and Madisetti explain that "peers do not [*sic*] need a trusted intermediary for interacting with each other," since a "blockchain network is not controlled by a central authority and all the transactions are verified and validated by a consensus among the peers" due to which "the peers do not need to trust each other." (Bahga and Madisetti 2016, p. 543).

Because of this trustless element, blockchains have been advocated in numerous applied contexts. For example, Banafa claims that "the decentralized, autonomous, and trustless capabilities of the blockchain make it an ideal component to become a foundational element of IoT [Internet of Things] solutions." (Banafa 2017, p. 2). Similarly, Kurtulmus and Daniel propose that "it is possible to create contracts that offer a reward in exchange for a trained machine learning model for a particular data set [which] would allow users to train machine learning models for a reward in a trustless manner," (Kurtulmus and Daniel 2018, p. 1802).

Along similar lines, Schaub et al. propose a reputation system for e-commerce using blockchain which would be trustless because of its privacy-preservation mechanism (Schaub et al. 2016). Another intriguing example is when Klems et al. propose the concept of *trustless intermediation* in the context of decentralized service marketplaces, claiming that "by leveraging blockchain-enabled smart contracts, we eliminate the need for trust in marketplace intermediaries and reduce barriers of entry, lock-in, and transaction costs," going as far as to say that they would be "removing now *obsolete* trust-establishing mechanisms," (2017, p. 731, emphasis added). Strong wording such as "obsolete" goes to highlight the degree to which conventional notions of trust appear to have been superseded by the novel promise of blockchain trustlessness. Yet as mentioned in the previous section, this reflects a conflation of the specific trustless trait of blockchain (for third-party verification), with broader considerations of "trust." The following sections present examples of how the fostering and nurturing of trust still remain necessary due to a dimension of human engagement. The human element is highlighted in four instances: direct interventions of "hard forks," the [mis]management of cryptoexchanges, the emission of initial coin offerings, and recourse sought by investors in traditional accountability institutions.

3 Hard Forks

The first example of human engagement which requires an element of trust is in hard forks, which are implemented to remedy or repair issues in the establishment of blocks on a cryptocurrency chain. Put simply, a hard fork is a radical change to the protocol that makes previously invalid blocks/transactions valid, and in that sense, represents a direct human intervention in blockchain construction (see Chohan 2017b, 2019a). Hard forks constitute a permanent divergence from a previous version of the blockchain, and so require all users (or "participant nodes") to upgrade to the latest version of a protocol software (see Destefanis et al. 2018). This is a trust-based activity in that all participants must voluntarily adapt to a remedied version and agree to continue along a new path, thereby agreeing to no longer accept another version of the blockchain.

The term "fork" is used because diagrammatically there is a split in the chain which resembles the prongs of a utensil, where one path follows the new and upgraded blockchain, while the other path continues along the older chain (Chohan 2017b). Meanwhile, the distinction of "hard" and "soft" refers to the means of splitting the blockchain: a hard fork creates two blockchains, and a soft fork is meant to result in but one chain.

The splitting of the path of a blockchain is accomplished through the deliberate invalidation of transactions confirmed by nodes that have not been upgraded to the new version of the protocol software. Human engagement lies in those participant nodes on the old chain coming to the realization that their version of the blockchain is outmoded, thus requiring their voluntary upgrade onto the latest version. For larger blockchain instruments, this tends to occur within a brief interval, but it is nonetheless a function of human decisions and interventions.

The most important reasons for the implementation of hard forks include: correcting important security risks found in older versions of the code; adding new functionalities; and reversing transactions (see Destefanis et al. 2018). A famous example of the former objective (remedial action against security risks) occurred during the hack on the DAO ("Decentralized Autonomous Organization," see Chohan 2017b). As Chohan recounts, nearly as soon as the DAO was launched, it became the victim of predatory attacks (Chohan forthcoming-b), which then necessitated human intervention to remedy the nearly $50 million worth of funds that were compromised. Following the hack, the blockchain Ethereum community voted unanimously in favour of a hard fork to undo the transactions which were responsible for the siphoning of the funds (denominated in tokens of the DAO). Technically, the voting mechanisms did not unwind the transaction history of the DAO, but instead relocated funds tied to the DAO to a new smart contract (see "smart contracts" in Chohan 2017b), which was programmed to allow the original token holders to withdraw them. Evidently, voting in this instance cannot be construed as anything but a human trust-based activity, while the process of overseeing the restoration of the funds of the DAO also constitutes a trust-based action (the trust being instilled in the curators to execute).

But whereas many hard- and soft-forks represent a consensus-based effort to remedy a problem in a blockchain system, not all such interventions are benevolent. The case of the exchange Quadriga CX, which shall be discussed in subsequent sections, also involved the insertion of a hard fork which, according to previous legal counsel, "started the company down a path of lawlessness," (Duhaime 2019). Hard forks thus represent one of the various human trust-based elements in the realm of cryptocurrencies, since assigned persons act deliberately to alter the process of blockchain formation towards a new direction.

4 Cryptocurrency Exchanges

Cryptocurrency exchanges (also abbreviated as portmanteau to "cryptoexchanges") are clearinghouses, traders, stores and/or market-makers for the sale and purchase of cryptocurrencies (Chohan 2018a). They can be described as "nodes for the transactions of crypto instruments between buyers and sellers [and the] juncture at which the human element becomes crucial," (Chohan 2019b, p. 3). Given the proliferation of cryptoinstruments and the widening public interest in ownership of cryptocurrencies, the number of cryptoexchanges had ballooned in the past few years (although much market consolidation is now taking place). As of this writing, it is estimated that there are more than 200 cryptoexchanges worldwide, amounting to a daily volume of nearly $15 billion US dollars across more than 8000 daily transaction pairs (Coinmarket 2019). Some of larger and more famous exchanges currently in operation include Binance (Maltese), Huobi (Singaporean), and Upbit (South Korean).

Cryptoexchanges may exchange pairs of cryptocurrencies or deal in fiat currencies, and can either charge bid-ask spreads when acting as market-makers or charge fees as matching platforms (Chohan 2018a). These exchanges occupy a position of centrality within the commercial ambit of cryptocurrencies. However, their mushrooming has not come without drawbacks, emanating largely from the unregulated nature of these spaces (Chohan 2018a, 2019b). Situated largely outside the regulatory space of traditional finance (see Chohan forthcoming-a), cryptocurrency exchanges suffer from substantial risks of theft, malpractice, fraud, and losses of abrupt shutdowns, and two examples are illustrative of this: Mt. Gox and Quadriga CX (Chohan 2019b).

4.1 Mt. Gox

Mt. Gox was a behemoth exchange located in Japan (but with mostly Western managers) which was handling up to 70% of the global Bitcoin trades during the period 2013–2014 (Chohan 2018a). In February 2014, Mt. Gox suspended trading, closed its website and services, and filed for bankruptcy protection from creditors, thereby spreading panic throughout the cryptocurrency markets. Two months later, the company began liquidation proceedings. It should not be lost on observers that human trust-based dealings were violated in the Mt. Gox episode, and it is important to see why.

At the time of the company's immediate crisis, Mt. Gox announced that approximately 850,000 bitcoins belonging to customers and to the company had gone missing, suggesting that they had been stolen (Chohan 2018a). This amounted to more than $450 million at the time. In the period since, 200,000 bitcoins have been "found," and the reasons for the disappearance: theft, fraud, mismanagement, or a combination thereof; have also been clarified.

Evidence presented in April, 2015 by the Tokyo-based security company WizSec concluded that "most or all of the missing bitcoins were stolen straight out of the Mt. Gox hot wallet over time, beginning in late 2011," but also that mismanagement and fraudulent practices had permeated the workings of Mt. Gox (Chohan 2018a, 2019b). An alleged internal crisis management document that was leaked had claimed that the company was insolvent, after having lost 744,408 bitcoins in a theft which went undetected for years (Chohan 2018a). With this, Mt. Gox's French CEO Mark Karpelès was arrested in August, 2015 by the Japanese police and charged with fraud and embezzlement, and manipulating the Mt. Gox computer system to increase the balance of a private account. Pursuant to interrogation, Japanese prosecutors accused him of misappropriating ¥315 M ($2.6 M) in Bitcoin deposited into their trading accounts by investors at Mt. Gox, and moving it into an account he controlled, approximately 6 months before Mt. Gox failed in early 2014.

4.2 *Quadriga CX*

A second disaster in the cryptoexchange category was embodied by the Canadian exchange Quadriga CX, which up until the death of its founder in late 2018, was the largest exchange in Canada by volume traded (Chohan 2019b). Quadriga CX initially took great steps to comply with regulation (Duhaime 2019), and was responsible for establishing the second Bitcoin automated-teller machine (ATM) in Vancouver, British Colombia in January, 2014. Rapid expansion, as well as the overall bullish trend in cryptocurrency prices, propelled Quadriga to a substantial position within the international cryptocurrency space by 2017. However, Quadriga's progress was marred by difficulties similar to those of Mt. Gox (see discussion in Chohan 2019b), including persistent difficulties in allowing customers to withdraw dollars in exchange for cryptocurrency.

At the centre of both the rise and fall of Quadriga was its founder, Gerald Cotten, who had used extreme cybersecurity and anti-hacking measures to protect the access to held cryptocurrencies. In December, 2018, Mr. Cotten passed away from complications relating to Crohn's disease while he was in India building an orphanage. As the key man behind the Quadriga CX exchange, only he had total access to the entirety of the exchange's holdings. As of January, 2019, that holding amount was equivalent to $190 million. Having rigorously protected the accessibility to the cryptoassets, Cotten's death forced the exchange to seek creditor protection, since it could not access its holdings to repay its counterparties. Accusations have been made, which are still under court proceedings as of this writing, that the underlying cryptocurrency assets may not have been present in the stipulated amounts—which is to say that the exchange did not have these holdings to repay counterparties in any case (Chohan 2019b), thereby suggesting fraud and theft.

Chohan (2019b) has highlighted that this creates the key man risk in the cryptocurrency space, which is usually associated with traditional financial firms that depend on the competency, power, or knowledge-base of a single individual. This is counterintuitive, given that cryptocurrencies purport to be decentralized, autonomous, and trustless systems. The risk of one individual's shortcomings, therefore, should not have a significant impact on the function of cryptocurrency systems, including exchanges. Yet the reality, as evidenced by Quadriga CX, is that these risks continue to loom large (Chohan 2019b).

Aside from the two major cryptoexchanges of Mt. Gox and Quadriga CX, there have been a slew of thefts, frauds, abrupt shutdowns, and regulatory actions against other exchanges (Chohan 2018a, b, c). Hong Kong-based Gatecoin was shut down due to failure to recover stolen funds in 2019, and the US Federal Bureau of Investigation (FBI) seized the assets of 1Broker, based in the Marshall Islands, for breaking a series of US security laws in 2018. The Australian financial intelligence agency AUSTRAC shut down two exchanges in the country after discovering links to illicit activities between the cryptoexchanges and organized crime rings in 2018. Regulatory authorities are also denying the applications of new exchanges for failing to meet standards relating to operations and compliance (Coin Asset Thailand is but one recent example).

Further still, the economic pressures on cryptocurrencies owing to weak demand and cheap production (mining) of cryptocurrencies have also led to closures at smaller exchanges (Liquid in the Ukraine being but one example). This in turn is fostering a consolidation in the cryptoexchange space, with larger exchanges such as Binance and Coinbase acquiring smaller players (Chohan 2018a).

All of this market activity is quintessentially human, not just in the illegal or dubious activities, but also in the categories of regular market behaviour (consolidations etc.). The trust-element is particularly important in this context as the malpractices of certain exchanges and investor-participants has had reputational consequences for cryptocurrencies as a whole. Those malpractices are premised on human drivers (greed, competitiveness) and so they sully the repute of cryptocurrencies, itself a non-traditional financial and technological space, in a manner that may jeopardize the viability of the space as a whole in the longer run. What is to be emphasized is that such activities are premised on inherently human traits, and so the conflation of third-party "trustlessness" with the trust requirements of cryptocurrency market behaviour is misleading.

5 Initial Coin Offerings

An Initial Coin Offering (ICO) is the mechanism by which capital is raised from investors through the emission of cryptocurrency coins (or "tokens," see Adhami et al. 2018; Chohan 2017d), usually in exchange for another cryptocurrency or for fiat money such as the United States dollar or the Euro (Fisch 2019; Chohan 2019a, b), and often expressed as a percentage of total newly issued currency (Catalini and Gans 2018). Adhami et al. describe ICOs as "open calls for funding promoted by organizations, companies, and entrepreneurs to raise money through cryptocurrencies, in exchange for a "token" that can be sold on the Internet or used in the future to obtain products or services and, at times, profits," (Adhami et al. 2018, p. 64). Therefore, ICOs can be seen as a new motor for raising investment capital (Howell et al. 2018; Lee et al. 2018; Adhami et al. 2018; Catalini and Gans 2018), and they offer "significant promise for new startups in the cryptocurrency space as means of quicker and easier capital raise," (Chohan 2017d, p. 3).

Yet there is a quintessential element of trust that is necessary for the emission of coins and their subsequent trading, and as this section of the chapter discusses, that trust has been compromised to quite an extent. As with cryptoexchanges, hard forks, and cryptocurrencies themselves, ICOs have mostly occurred in the online realm that lies beyond regularized and traditional finance, devoid of the structures of financial regulation which allow for contemporary capitalism to function in a more lawful and stable manner (Fisch 2019; Howell et al. 2018; Chohan 2017b, 2019a, b).

ICOs are "bypassing any regulation that normally applies to businesses placing securities to retail investors, [and so] dozens of developer teams and entrepreneurs collect money in absence of official prospectuses, with no particular protection for contributors and disclosing only a very limited set of information," (Adhami et al. 2018, p. 65). Furthermore, "there are many scams, jokes, and tokens that have nothing to do with a new product or business," (Howell et al. 2018, p. 1).

To this point, ICOs have "low contributor protection, a limited set of available information, [almost] no supervision by public authorities, and [almost] no relevant track record for proponents," (Adhami et al. 2018, p. 73). Benedetti and Kostovetsky (2018) have surmised that ICOs are in fact a digital reiteration of the *Tulip Mania* which engulfed Europe in the early decades of the seventeenth century. Large numbers of ICOs have resulted in "substantial scam-artistry, phishing, Ponzi schemes, and other shenanigans" (Chohan 2017d, p. 5). According to one study which examined the lifecycle of ICOs from the initial proposal to the final phase of trading on a crypto-exchange, more than 80% of ICOs emitted in 2017 were scams (Satis Group 2017), amounting in value terms to more than $1 billion US dollars (value estimates of the total capital raised in that year was $11 billion). For 2018, another ICO advisory firm estimated that, for more $20 billion in capital raised from 789 ICOs, the 10 largest ICO scams swindled a combined amount of more than $700 million (Fortune Jack 2018). Benedetti and Kostoyevsky have determined that only 44.2% of startups survive after 120 days from the end of their ICOs (Benedetti and Kostovetsky 2018).

Much of this is premised on the breach of trust in the cryptocurrency domain due to lackadaisical levels of investor due diligence, the wildly inflated promises of transformation made by issuers, and the quintessential human traits of greed and "fear of missing out" (colloquially termed "FOMO"). Humans are the ones exerting agency, irrespective of the lack of third-party verification that characterizes blockchain technologies. Indeed, investors who have dealt with dubious ICOs have fallen prey to the seemingly endless rhetorical promises of the cryptocurrency realm, but there was far less complaint about ICOs when cryptocurrency prices were at their zenith (see Chohan 2018a, b, c). Rather, it was when the prices declined that the furore of losing investors spread across the online forums and into the public sphere.

As with cryptoexchanges, the scope of widespread financial abuse through ICOs came to jeopardize the reputation of the space as a whole (Chohan 2019a, b), with many small- and large-scale investors demanding recourse and recovery of funds. The dilemma that this has posed for recourse to traditional (human-led) institutions of regulation is discussed in the next section.

6 Recourse of Traditional Structures

Given that the inherent design of cryptocurrencies is to situate them outside the traditional financial architecture (Fisch 2019; Howell et al. 2018), any demand by investors for financial recourse from traditional institutions poses a dilemma for regulatory authorities around the world. Initially, the dilemma sprung from the sheer bewilderment at the meteoric rise of the sector (see Chohan 2017b, 2018c), but since then the dilemma has been spurred by the need to strike a balance between fostering innovation and imposing accountability (see Chohan forthcoming a, b). The Securities and Exchange Commission (SEC) and Commodities Futures Trading Commission (CFTC) have framed that balance as the need for regulations to "set and enforce rules that foster innovation, while promoting market integrity and confidence," (Clayton and Giancarlo 2018).

After all, it is humans who have wronged other humans, only to seek remedial action from yet other humans for those wrongs. A sense of trust has been breached, and now the onus is on regulatory authorities to restore trust. Two instances are illustrative of this problem, (1) the SEC and CFTC regulation on ICOs, and (2) the aforementioned case of Quadriga CX exchange.

6.1 SEC and CFTC as Regulators

First, the onus for the restoration of trust has fallen on regulators, particularly American authorities, who have had the richest experience with the regulation of cryptocurrencies thus far. Specifically, action has come from the SEC and CFTC in the United States (Chohan forthcoming-a). As far back as 2017, the Chairman of the SEC had insisted upon investors that there was a need for investors to exercise greater caution, given the possible breaches of trust and the financial dangers of being misled by fraudulent cryptocurrency agents (Clayton 2017). He also demanded that laws "provide that investors deserve to know what they are investing in and the relevant risks involved," (Clayton 2017).

The SEC chairman then declared that the commission's Division of Enforcement would "police this area vigorously and recommend enforcement actions against those that conduct initial coin offerings in violation of the federal securities laws," (Clayton 2017). These are the sorts of breaches of trust that regulatory authorities have warned of and also taken action against. Chohan notes that a substantial series of enforcement actions have since been taken against ICO issuers and cryptoexchanges who have not complied with securities regulation (Chohan forthcoming-a), a few examples of which have been mentioned in previous sections of this chapter.

In June, 2018, a joint statement was issued by the chairmen of the SEC and CFTC (see Clayton and Giancarlo 2018) which emphasized closer cooperation between their agencies while insisting upon the need for regulations to strike a balance between fostering innovation on one hand and promoting market integrity and confidence on the other. Both of these regulatory bodies are setting the trend for international regulators in protecting investors and regularizing ICOs and cryptoexchanges, particularly since a growing public pressure in the wake of volatile (and declining) prices of cryptocurrencies and a massive scale of fraudulent activity has fomented investor anger and the desire for recourse in traditional institutions.

6.2 *Quadriga CX in Court*

The aforementioned Quadriga CX case, with the death of its CEO Gerald Cotton, also embodies an example of both (1) the key man risk, which is a human problem of entrusting an individual with excess knowledge or power (Chohan 2019b), and (2) the trend of investors seeking recourse in human institutions. As mentioned previously, upon the announcement of the Mr. Cotten's untimely demise in India while on a humanitarian trip, a frantic reaction pervaded the cryptocurrency markets, This is because the underlying amounts were by no means trifling: up to US$190 million owed to perhaps 115,000 customers has been missing or cannot be accessed as of this writing, because only Mr. Cotten held the password to off-line cold wallets (Chohan 2019b).

The inaccessibility to the cryptocurrencies thus incited creditors of the company to take Quadriga CX to a source of traditional recourse—a court in Halifax, Nova Scotia, Canada. In that jurisdiction, Quadriga CX has been granted, as of this writing, a stay order under the temporary legal protection from its creditors under the Companies' Creditors Arrangement Act, a form of bankruptcy protection instated during the Great Depression to prevent firms from falling into total insolvency. That stay order will remain in effect until the middle of 2019, but may be extended thereafter. It has yet not been determined if foul-play is involved, in the manner that the aforementioned Mt. Gox case in Japan came to reveal.

What is of more pressing concern is that the trust-based element in the Quadriga CX case has been breached, and the remedial action is being sought in the traditional domain (a Canadian court). Since Mr. Cotten had not instated any mechanism for access to his off-line cold wallets in a situation of extended actuarial absence (death or disability), the cryptocurrency may be missing, lost, or irretrievable. An element of trust between the parties has been violated, and the traditional judicial system has been brought to bear on the matter.

7 Conclusion

This chapter has sought to present a nuance around the concept of "trustlessness." While it is indeed true that blockchain technologies do not require third-party verification and are therefore trustless in one sense, the conflation of the term with other notions of trust has been an unproductive one. Trust is basis of all sustained commercial life, and to suggest that the older trust-establishing mechanisms are "obsolete" (Klems et al. 2017, p. 731) is misleading, as four different forms of examples in this chapter show. Indeed, the problem of human trust persists in the domain of cryptocurrencies, and has been summarized thus:

> Much of the discourse on cryptocurrencies has sought to detach it from problems that beset the domain of traditional finance. This is somewhat misleading, for while the substance of cryptocurrencies themselves may be congruent with the decentralized, trustless, autonomous, immutable principles championed by cryptoanarchism, the praxis of cryptocurrency transactions still contains a significant human element, including that of engaging in transactional activity (Chohan 2019b, p. 3).

The notion of exaggerating "trustlessness" stems in part from inaccurate marketing, but also in part from the philosophical underpinnings of cryptocurrencies, which lie within *cryptoanarchist thought*. Cryptoanarchism seeks to cultivate decentralized, autonomous, and voluntary exchange among individuals in a manner that protects their identities, and therefore their risk of persecution, from structures of established authority (Chohan 2017e). This creates an ambiguity as to the level of trust that would be forthcoming in the course of regular engagement among participants.

For cryptoanarchism, as with anarchism itself (see Wolff 1998; Marshall 2009), there are utopian expectations of human beings that remain wanting, including a selflessness and trust between groups of people who will demonstrate respect and consideration in an effort to come to mutual aid. In an anonymous world of trading bits of code as monetized instrument, even as it may be nominally "trustless," issues of trust have indeed surfaced, and often bitterly so.

Sometimes the fault has been exogenous, as with the hacks and external attacks on the security of networks, exchanges, and tokens. At other times, however, the problem has been endogenous, as with the thefts, misrepresentations, Ponzi schemes, and frauds that have been perpetrated across the cryptocurrency space, whether on cryptoexchanges, coin offerings, or on networks themselves.

The principle assertion of this chapter then becomes that fuller clarity is required in the universe of cryptocurrency participants as to what trustlessness really means. Third-party verification aside, there is indeed a dire need for strengthening trust in a realm that is both digital and largely anonymous. Regulation and oversight are natural mechanisms for helping to ensure this, but there are many limitations to the implementation of cryptocurrency regulation, accountability and enforcement, not least at the international level. Indeed, it is inherent to the very design of cryptocurrencies that they should help to mask and protect the identities of participants. But the failures of such systems, as indicated throughout this chapter, thus raise questions about the viability of cryptocurrencies as complementary (or to the idealists: parallel) monetary systems in the longer-run. A balance too must be struck between fostering continued innovation and insisting upon accountability. These questions must be more fervently explored in future research, as well as in policy praxis, as cryptocurrencies shed the perception of being disruptive shadowy technologies, and begin to collectively come-of-age as mature technological and financial instruments.

References

Adhami S, Giudici G, Martinazzi S (2018) Why do businesses go crypto? An empirical analysis of initial coin offerings. J Econ Bus 100:64–75

Bahga A, Madisetti VK (2016) Blockchain platform for industrial internet of things. J Softw Eng Appl 9(10):533

Banafa A (2017) IoT and blockchain convergence: benefits and challenges. IEEE Internet of Things

Benedetti H, Kostovetsky L (2018) Digital tulips? Returns to investors in initial coin offerings. Returns to investors in initial coin offerings (May 20, 2018)

Catalini C, Gans JS (2018) Initial coin offerings and the value of crypto tokens (no. w24418). National Bureau of Economic Research

Chohan UW (2017a) Cryptocurrencies: a brief thematic review. Notes on the 21st century. Critical Blockchain Research Initiative (CBRI) Working Papers. https://papers.ssrn.com/sol3/papers.cfm?abstract_id=3024330

Chohan UW (2017b) The decentralized autonomous organization and governance issues. Notes on the 21st century. Critical Blockchain Research Initiative (CBRI) Working Papers. https://www.ssrn.com/abstract=3082055

Chohan UW (2017c) The double spending problem and cryptocurrencies. Notes on the 21st century. Critical Blockchain Research Initiative (CBRI) Working Papers. https://papers.ssrn.com/sol3/papers.cfm?abstract_id=3090174

Chohan UW (2017d) Initial coin offerings (ICOs): risks, regulation, and accountability. Notes on the 21st century. Critical Blockchain Research Initiative (CBRI) Working Papers

Chohan UW (2017e) Cryptoanarchism and cryptocurrencies. Notes on the 21st century. Critical Blockchain Research Initiative (CBRI) Working Papers. https://www.ssrn.com/abstract=3079241

Chohan UW (2018a) The problems of cryptocurrency thefts and exchange shutdowns. Available via SSRN 3131702

Chohan UW (2018b) Bitconnect and cryptocurrency accountability. Notes on the 21st century. Critical Blockchain Research Initiative (CBRI) Working Papers. https://papers.ssrn.com/sol3/papers.cfm?abstract_id=3131512

Chohan UW (2018c) Oversight and regulation of cryptocurrencies: bitlicense. Notes on the 21st century. Critical Blockchain Research Initiative (CBRI) Working Papers. https://papers.ssrn.com/sol3/papers.cfm?abstract_id=3133342

Chohan UW (2019a) Are cryptocurrencies truly 'trustless'? Notes on the 21st century. Critical Blockchain Research Initiative (CBRI) Working Papers. https://papers.ssrn.com/sol3/papers.cfm?abstract_id=3331544

Chohan UW (2019b) The key man problem in cryptocurrencies? Case of QuadrigaCX. Notes on the 21st century. Critical Blockchain Research Initiative (CBRI) Working Papers. https://papers.ssrn.com/sol3/papers.cfm?abstract_id=3329573

Chohan UW (forthcoming-a) Public value and citizen-driven digital innovation. Information Polity

Chohan UW (forthcoming-b) Public value without public managers? Decentralized autonomous organizations and public administration. Information Polity

Clayton J (2017) Statement on cryptocurrencies and initial coin offerings. In: Securities and exchange commission (SEC) statements. SEC, Washington, DC

Clayton J, Giancarlo JC (2018) Joint statement on cryptocurrencies and initial coin offerings. Securities and Exchange Commission (SEC) & Commodity Futures Trading Commission (CFTC), Washington, DC

Coin Market (2019) Top cryptocurrency exchanges list. https://coin.market/exchanges. Accessed 20 April 2019

Destefanis G, Marchesi M, Ortu M, Tonelli R, Bracciali A, Hierons R (2018, March) Smart contracts vulnerabilities: a call for blockchain software engineering? In: 2018 international workshop on blockchain oriented software engineering (IWBOSE), IEEE, pp 19–25

Duhaime C (2019) From law to lawlessness: bits of the untold QuadrigaCX story. Coindesk. March 26

Fisch C (2019) Initial coin offerings (ICOs) to finance new ventures. J Bus Ventur 34(1):1–22

Foroglou G, Tsilidou AL (2015, May) Further applications of the blockchain. In: 12th student conference on managerial science and technology

Fortune Jack (2018) Study on ICO scams in 2018. https://fortunejack.com/

Howell ST, Niessner M, Yermack D (2018) Initial coin offerings: financing growth with cryptocurrency token sales (no. w24774). National Bureau of Economic Research

Kiviat TI (2015) Beyond bitcoin: issues in regulating blockchain transactions. Duke Law J 65:569

Klems M, Eberhardt J, Tai S, Härtlein S, Buchholz S, Tidjani A (2017, November) Trustless intermediation in blockchain-based decentralized service marketplaces. In: International conference on service-oriented computing. Springer, Cham, pp 731–739

Kurtulmus AB, Daniel K (2018) Trustless machine learning contracts; evaluating and exchanging machine learning models on the ethereum blockchain. arXiv preprint arXiv:1802.10185

Lee J, Li T, Shin D (2018) The wisdom of crowds and information cascades in FinTech: evidence from initial coin offerings

Marshall P (2009) Demanding the impossible: a history of anarchism. PM Press, Oakland, CA

Satis Group (2017) Study on ICO scams in 2017. http://satisgroup.io

Schaub A, Bazin R, Hasan O, Brunie L (2016, May) A trustless privacy-preserving reputation system. In: IFIP international conference on ICT systems security and privacy protection. Springer, Cham, pp 398–411

Wolff RP (1998) In defense of anarchism. University of California Press, Berkeley, CA

Blockchain and Alternative Sources of Financing

Othalia Doe-Bruce

Abstract For the past decade, Bitcoin has captured the attention of the world with its extreme price swings and yet constant price rises, outperforming most traditional asset classes. Along with Bitcoin came the phenomenon of the ICOs or Token Offerings, equally captivating, that turned start-ups' founders into overnight millionaires. For the first time, start-ups with a compelling vision, no longer needed to plead their case, by pitching to multiple stone faced VCs, all so they could raise a few hundred grants. Start-ups could now pitch to the world and raise millions in a matter of minutes. Prompting the CEO of one such start-ups that raised USD25 million and based in Toronto, Canada, to say: "The crypto world is not real, it is a dream". In this chapter, we will explore the various alternative sources of financing such as ICOs, STOs and IEOs that came along with the Bitcoin, their implications and what we can expect going forward.

1 Introduction

I like to think of Blockchain as a multi-purpose technology that is living and breathing unlike any other emerging technologies such as Artificial Intelligence or the Internet of Things. It is a technology that can be manned to retain the cold nature of the machine, remaining in the background of a large and inaccessible corporation, or a technology that can be transformed to create entirely new economies, open sourced, disrupting the way we interact with each other.

One example of such transformation was ushered in through the advent of cryptocurrencies when in 2009, Satoshi Nakamoto officially launched the first ever, completely secured and autonomous digital currency, backed by no governments, and only backed by the will of the people called bitcoin[1], Bitcoin (2019). With bitcoin[1], people, all over the world could transact value with each other, regardless of geographical limitations, national identities, or economic status.

O. Doe-Bruce (✉)
InnovFin Consulting Inc., Toronto, ON, Canada
e-mail: o.doebruce@innovfin.ca

S. Goutte et al. (eds.), *Cryptofinance and Mechanisms of Exchange*, Contributions to Management Science, https://doi.org/10.1007/978-3-030-30738-7_6

First, a Quick Overview of the Basics

Blockchain is the technology behind many distributed and decentralized applications. Cryptocurrencies are one such applications. Some Blockchains are permissioned or private, others are permissionless or public. Cryptocurrencies fall under the permissionless or public blockchain umbrella. All applications, built on top of a blockchain platform are called DApps for Decentralized Applications. Some prefer to use the term DApps or dapps. There are heated online debates as for which term is the "ONE" to be used as the norm. Unless agreeing on the "ONE" could help us solve world hunger, it is not a subject worth dwelling on.

2 The Ethereum Blockchain and the Ether Crypto-currency

Close to 5 years after the first cryptocurrency, bitcoin, was launched, a young Canadian, in the name of Vitalik Buterin took Satoshi Nakamoto's concept further creating the Ethereum[1] Blockchain and Ether[2] cryptocurrency, (Ethereum Foundation 2019). Vitalik, understood that Bitcoin as a blockchain platform was too self-limiting to cryptocurrency, the bitcoin[3]. Vitalik wanted to give the people a novel approach to transacting not only cryptocurrencies but also other digitizable values such as real estate, commodities, music rights and more.

These digital values will be called tokens, since unlike cryptocurrencies which aim to replace our traditional, physical or fiat money, tokens, served two purposes: (1) give the holder access to a specific blockchain platform or/and (2) become the digital representation of a physical asset. Later, the implementation of the Ethereum blockchain will give birth to very interesting tokenized projects, with some of my favorites and some currently functional (Noor 2019), including:

2.1 CryptoKitties

A platform to collect, buy and breed non-replicable, non-destroyable and uniquely digitized cats using the principles of token non-fungibility[4] and asset rarity. In 2017,

[1]"Ethereum" typically refers to the blockchain platform, that supports the "ether" cryptocurrency. In this chapter, we will often use "Ethereum" and "ether" interchangeably. The trading ticker of "ether" is ETH which will also be often used.

[2]Ibid.

[3]"Bitcoin" with a capital "B" typically refers to the blockchain technology platform, that supports the "bitcoin" cryptocurrency with lower case "b". In this chapter we will often use "Bitcoin" and "bitcoin" interchangeably. The trading ticker of "bitcoin" is BTC which will also be often used.

[4]Token Fungibility vs. Non-Fungibility: Fungibility is defined by Investopedia as an asset that can be interchanged with other assets or goods of the same type, (Laura Green 2019). For example, money is a fungible asset because it can be divided and subdivided in a any number of parts

one digital cat was selling for as high as USD100,000 a piece, with a total cats' sale of USD6 million by year end. Non-fungible tokens on the Ethereum platform are based off the ERC721[5] token standard. In parenthesis, on the Ethereum blockchain network, most tokens use the ERC20[6] token and are fungible, (Wikipedia, ERC-20 2019).

As of May 7, 2019, there were 185,387 ERC20 tokens on the Ethereum network. There are 64 types of ERC721 tokens in circulation as documented by Bloxy in 2019, (Bloxy 2019). Most were created for gaming purposes. Cryptokitties (AxiomZen 2019) is a creation of the award-winning venture studio, Axiom Zen.

The concept and price of cryptokitties, (Cryptokitties 2019) might sound futile, but it is interesting to imagine the potential applications of this form of digitization to physical assets that should not be replicated such as land or art. It is this concept that the *Arcona Digital Land* somewhat aimed to realize with their version of ERC721 tokens (Arcona 2019). The Arcona project combines Augmented Reality, ERC721 tokens and a gaming environment to allow users to build, own, sell and buy virtual worlds, using the parameters of real-world pieces of lands.

2.2 Brave Browser

With its Basic Attention Token (BAT) (2019), the Brave browser is a personal favorite dApp of mine, downloadable on Windows, Mac OS and Linux. This browser automatically removes cookies, digital ad pop-ups and more from any website visited by a user. As a result, the web surfer's device consumes less electricity and data, and therefore have a longer battery life while saving cost. The user, however, can elect to watch any number of ads, and is rewarded for her "Attention" using the BAT token.

The Brave browser is a proudly blockchain based app, graced to the world by the founders of JavaScript and the co-founder of Mozilla and Firefox.

2.3 uPort

Another favorite of mine, uPort opens a new world of opportunities for *digital identities*, by giving identity ownership back to the individual. In other words, it

---

representing the same asset, (District0x Educational Portal 2019). A real estate property on the other hand is not interchangeable, in its physical form that is, as it is unique and indivisible. On the Ethereum network, ERC20 tokens have a fungible standard while ERC721 tokens follow a non-fungible standard.

[5]Ibid.

[6]Ibid.

enables the *Self-Sovereign Identity*. On uPort, you can create an identity on the Ethereum blockchain network, log-in securely to applications without passwords, manage your personal information and verifications, approve Ethereum transactions and digitally sign files, (uPort 2018).

You can imagine how the use of this App could apply to larger scale projects such as the Future of Smart Cities. In that regards, using your uPort ID, you can test and play this future application in the virtual demo city of *uPortlandia*. The vision is to use one single uPort to access several services such as: Public Services, Diploma & Employment Verification, Insurance Coverage and more. Very exciting stuff if you ask me. You can download the app for iOS and Android. uPort is part of the Consensys Formation. Consensys, founded by Joseph Lubin, a co-founder of Ethereum, is a global formation of technologists and entrepreneurs working to enable a decentralized world by building decentralized applications on Ethereum.

Throughout this chapter we will use cryptocurrencies, tokens and coins interchangeably to designate digital value being transferred on top of a permissionless or public blockchain platform.

3 A New Age of Financing, Birth of the ICOs

Now that all kind of digital assets could be created and transacted thanks to the invention of *Vitalik Buterin*, the people to which the gift was offered, saw in the novelty an alternative way of accessing financing. A new way to give the middle finger to the arduous long and often disadvantageous process of raising funds through Equity and Venture Capitalists, a new source of alternative financing called the ICO. It is important to note that while the Ethereum Blockchain launched the wave of tokens and ICOs, other blockchain platforms such as *Cosmos, EOS, Tezos, Augur, Aion, Neo* and the *Stellar Network* soon followed suit; each adding their own flavor to the *native coin*[7] pot. Regardless, in the world of public blockchains, BTC and ETC hold the two largest market caps and therefore, merit, in my opinion the most attention.

3.1 What is an ICO?

An ICO, defined as an Initial Coin Offering, is a novel way of raising financing using tokens or cryptocurrencies, without necessarily releasing any equity to the token

[7]Native coins designate cryptocurrencies or tokens that are integral part of a blockchain platform, the protocol layer. Native to the very blockchain upon which dApp and their tokens will be created. Think Ether the native coin, BAT the dApp token build on top of the Ethereum blockchain.

buyers or investors. Essentially, a company looking to build a blockchain platform and launch an ICO will use the following steps:

1. Issue a whitepaper[8] outlining the vision and purpose of the proposed blockchain project. The whitepaper will contain details such as: Project Mission, High Level System Functions & Architecture, Use Cases, Tokenomics[10], Token Distribution, Team Background & Bio, Advisors & Partners, Project Roadmap and Timeline.

   Whitepapers (Wikipedia, Whitepaper 2019) are marketing tools destined to gain the buy in of investors and/or future customers. In some cases, the company will also issue a yellowpaper[9] (Melanie Clay 2018; WikiCryptoCoins 2018), a deep dive into the often-unproven technological specifications the project intends to utilize. Often than not, yellow papers are a work in progress and can experience many updates along the way, until the blockchain platform reaches full implementation.

   Yellow papers are not primary requirements for the purpose of ICO fundraising. In researching these for yourself, you will find any number of papers whether white, yellow, beige (Jerry Yu 2018) or otherwise, some with more details and technical information than others.

   The most important principle to keep in mind, is that anyone looking to invest in a blockchain project should (1) gather as much information as possible on the project, in and out of paper, (2) read the gathered information thoroughly to the best of your ability, (3) and most importantly understand the project. If you do not understand the proposed blockchain solution, I'd recommend you put your hard earn money towards other purposes. If you do blindly invest, you would be speculating, and putting your faith in the hands of the god of luck and providence.
2. Raise awareness about the project and whitepaper through an aggressive Marketing, Public Relations and Social Media strategy in order to gather interest from the public and investors.
3. Build a community of supporters who will hold the team accountable to their promised results and challenge some of the ideas listed in the whitepaper.
4. Whitelist potential investors with interest in purchasing the tokens or cryptocurrencies. Investors selection will depend on the strategic direction undertaken by the team.

   For example, the team might decide to only allocate tokens to investors with a minimum of USD50,000* and above. As another option, the team might decide to allocate tokens to any amount of investment, with an affordable floor as little as $20 and a cap of USD5000. Alternatively, the team might also decide to restrict investments to institutional investors with a minimum of investment of

[8]A white paper is an authoritative report or guide that informs readers concisely about a complex issue and presents the issuing body's philosophy on the matter, (Jerry Yu 2018). It is meant to help readers understand an issue, solve a problem, or decide.

[9]A yellow paper is a document containing research that has not yet been formally accepted or published in an academic journal.

USD100,000 to more. We will further expand on the involvement of institutional investors in the ICO market later in this reading.

5. Set up the smart contract that will be utilized to accept investors' funds and release the newly minted tokens to the investors, at a future agreed upon date, according to the Tokenomics[10] (Rajarshi Mitra 2018) and timeline outlined in the whitepaper. Let's take a few minutes to grasp the implications of this sentence.

   The ICO market up until the end of 2017 had been highly unregulated. This meant that any Jane and Joe working out of their parents' basement, could release a whitepaper, with no legal or regulatory oversight, promising the sky, the moon and everything afterward, to a group of remote supporters or community. If Jane and Joe were convincing enough, they would receive the buy in of their community and set up a *smart contract*, which will in turn be used to collect investments.

   If you are not familiar with this term, *smart contracts* are codified contracts created to enable the automated reception of crypto-currency payments prior to the construction of a blockchain platform. The automation is set up as per the tokenomics[10] and guidelines of the accompanying whitepaper. The whitepaper, however, informative, is not legally binding.

   Investors in ICOs are therefore only relying on the words of the team behind the project, to bring the project to fruition. Often, ICOs were erroneously likened to stocks which was partially the case. To clear the air, like stocks, ICOs are an alternative way of raising funds from the public and that is where the similarity stops.

   Unlike stocks, ICOs do not give the holder the right to any equity or revenue shares of the company behind the project. ICOs only give an investor access to the token he purchased, with no obligation whatsoever from the company to build the said platform in the future or work to increase the price of the coin. Meaning that, from the very beginning ICOs had their work cut out for them, when it came to establishing trust between team and communities.

   It also meant ICOs were too tempting of an opportunity for malicious individuals to not want to commit fraud. However, I digress, more on that later.

6. To give reassurance to the community behind a project, pre-ICO, many companies will typically have their smart contract audited by a third-party, well-known and established company such as Hosho (in this infant industry establishment is relative).
7. KYC[11] the selected investors to establish a certain level of regulatory compliance in the fundraising process. KYC however, had not been much of a concern for most of the ICOs launched up prior to 2017 as there was a very limited, to no

[10]Tokenomics is a set of rules, principles and incentivization mechanism that govern a crypto-currency or a token ecosystem, with the goals of sustaining the vision of a blockchain based platform.

[11]KYC or Know your customer, is used to link the movement of funds with its sources, in order to prevent, illegal transactions such as money laundering or other fraudulent activities.

regulatory scrutiny. ICOs at the time were innovations that only early adopters engaged in, while most regulators took an observer stance in order to determine their implications and impact on the economy.

8. In 2017, the wave of post-ICOs failures with its fair share of what turned out to be ICO scams, and the 2017 crypto-recession or crypto-winter that will ensue will change that passive stance forever. Regulators in 2018 will switch gear and take a more aggressive approach with increased scrutiny as per the legality of ICOs.
9. Finally, on the date and time set for the ICO, and as per the whitepaper, fundraising will commence and investors all over the world will send cryptocurrencies (ETH or BTC for most) to a designated public address. In return for providing funding, investors will automatically receive the ERC20 tokens associated with the project, or in some cases receive tokens at a much later date, if subject to a lock-up period.
10. It is important to note that ERC20 tokens can be sent temporarily or permanently to investors depending on whether the project is built for a native coin or not.
11. If native an additional step will involve, waiting until the production platform has been launched to have your previously issued tokens burnt, and exchanged for the newly completed cryptocurrency.
12. Essentially this last step will conclude the process of fund raising via ICOs.
13. However, to receive full satisfaction from investors post-ICO, the team would need to also take on the following additional steps:
14. List tokens on crypto exchanges, with top tier exchanges that have substantial trading volume and liquidity such as Binance, Huobi, Coinbase, Coinsquare, OKEx, Einstein, Kraken to list a few in no particular order.

3.2 Cryptoexchanges

Top tier exchanges are typically very picky with accepting any new coin on their exchange due to the higher risk of scam projects in the unregulated cryptocurrencies. *Therefore, top tiers will engage in a very rigorous and stringent due diligence and background check of the team and project.* It is done so that the exchange's reputation and operations are not compromised by onboarding a project with a weak or potentially fraudulent team. It is without saying that onboarding a new cryptocurrency or token on an exchange requires development efforts, in addition to trading efforts including assigning market makers to the new crypto currencies to ensure the liquidity of the coin.

In parenthesis, contrary to traditional century old stocks, fixed income, commodity and other exchanges such as the New York Stock Exchange, the Toronto Stock Exchange, the CBOX and others, market making in the crypto world is rather decentralized. If there is a need, anyone with the will to take on a challenge could reach out to an exchange to take on the role of a market maker and earn a percentage of the spread on traded coins. Blockchain, as I like to think, took decentralization to a whole new level of understanding, opening opportunities only previously accessible to the wealthy, and bringing it to the people (Jane and Joe). A blessing and sometimes a curse. Some lower tier exchanges accept any and all projects.

In the world of crypto-currency trading, there are two types of exchanges:

1. Exchanges that accept crypto to crypto trades, in which case all transactions between buyers and sellers are purely digitized. KYC are kept to a minimum as trades between different coins are considered exchanges of commodity.
2. Exchanges that would facilitate crypto to fiat trading, in which case the user interface allows the trader to send fiat wire transfers to the exchange using electronic forms of payments such as credit card, wire transfers, or money order to list a few; or convert crypto to fiat and withdraw fiat via cash, cheque or direct bank transfer.

   As expected, these exchanges have the added responsibility of interacting with traditional payment processors such as Visa and Banks. That interaction necessitates therefore, a more stringent and rigorous due diligence & KYC process of not only the project and the team behind the project, but also of the users of the exchange as imposed by the banks.

Listing coins on an exchange, is an arduous process that may come often months and months after the conclusion of the ICO. It is also a costly process that proceeds from the fundraise are partially used for.

3.3 Similarities Between ICO and Traditional Stock Market

Once the team has announced the exchange listing, investors are provided the go ahead to start trading the purchased coins. Typically, the investors would want to exit their investments, and leaving it up to late comers to buy the token from them at a higher price. In fact, ICO coins are sold in a staggered manner, at a discounted price to early investors who can then reap a higher profit once the coins hit an exchange. Early investors are therefore often first to dump coins, unless subject to a lock-up period as mentioned earlier.

I would like to highlight the similarity here with the traditional stock IPO, Initial Public Offering (Corporate Finance Institute 2019). Traditionally, in an IPO, early investors, typically institutions, enter early agreements to purchase the stocks at a discount, reaping the benefit of a large spread once sold to the public at a higher price. In crypto, early agreements can involve institutions as well as Jane and Joe. In the months that follow the ICO, the team behind the project will switch gears, shifting from focusing on grabbing investors' attention, to building the blockchain platform that was promised.

For many scam ICOs, or ICOs with weak and inexperienced team, this is where much of the effort stops.

After pocketing investors investments, with no much guarantee but their words, the project will dwindle out of existence. Some teams will simply disappear out of thin air or be found sipping champagne on a remote island. Despite these black sheeps, however, many projects are strong enough to persist, because the teams care enough about their reputation, so as not to disappear, or "fake their own death" as it has been the case. There are currently about 2215 crypto-currencies and tokens being traded on various crypto exchanges as reported by CoinMarketCap. Some are worth as little as a fraction of a cent.

There is a major difference between ICOs and the traditional IPO market, where platforms and products are built first, before calling out for public investments. In the ICO market, often, products and platforms are built second. After the 2018 crypto winter that wiped out close to 2/3 of the total global market cap, ICO investors turned cautious, and demanded that crypto projects show better advancements with their platform, and concrete and positive test results with a clear path to revenue post-fundraising. In addition, just as in the traditional market, there are the large caps, most stable, most traded coins, and the penny stocks which in crypto world have earned themselves, the term "shit coins".

3.4 *The Pump and Dump[12] Online Community*

There is much more to be said about the inner workings of the pre and post ICO market, why some investors risked investing in a coin and got burned, or got handsomely rewarded. Online "Pump and Dump[13]" groups, for example, are online communities that gather every once so often to pump the price of a shit to small cap coin, on social media, only to drop it like its hot as soon as the price has appreciated. Pumps do not limit themselves to small cap coins. There have been pumps of larger cap coins such as Bitcoin Cash in the past.

Pump and Dump online channels exist, mostly on Telegram, where communities of anonymous users, orchestrate the sudden increase and crash of coins, in order to reap profit by selling at the top of the pump. Most popular channels include The Big Pump Signal, Donald Pump, The Mega Pump Group and more.

For the naïve investor looking to "buy and hold" this is usually death. Pumps might go on for days, when a coin with no obvious reason begin appreciating very quickly, sometimes reaching 100× only to drop as quickly as it shot up. I like to assimilate the effect of a pump and dump to the shape of a camel back. The "camel back effect".

In the traditional stock market, this is market manipulation, a century old problem plaguing investment markets. Heavy regulations with time were put in place to

[12]Tokenomics is a set of rules, principles and incentivization mechanism that govern a crypto-currency or a token ecosystem, with the goals of sustaining the vision of a blockchain based platform.

[13]Ibid.

punish offenders engaging in activities that eroded investors' confidence in the market and are ultimately detrimental to the economy and the society.

Players with enough power and access to manipulate traditional stock markets are however not your average Jane and Joe. They are traders, investment managers or other high-profile finance professionals, working at institutions with assets' size big enough to move the market. Despite all regulations, stock market manipulations are still a very current issue. Take the case of former J.P. Morgan trader John Edmonds, whom as of this writing, is still awaiting his day in court for working with other co-conspirators to manipulate the price of precious metal markets, (Dawn Giel 2019).

In the crypto market, the same behaviour, is decentralized, meaning perpetrated by groups of Janes and Joes, hiding behind the relative anonymity enabled by the blockchain technology to manipulate the crypto market in the hopes of turning in high profit very quickly. This behaviour for now, escape from the grasp of the same regulations that somewhat control stock market manipulations.

3.4.1 Consequences of Pumps and Dumps

Until regulators catch up to the crypto market in general, these activities will continue adding to the riskiness of an already risky crypto market. Most sadly, this behaviour negatively tint an otherwise promising industry.

As a member of the CFA Institute, an organization that aims to uphold the most ethical practices of investment management for the benefit of society, "Pumps and Dumps" in the crypto market make my stomach turn. Mostly because like in the stock market, they do not promote a healthy crypto economy, which ultimately prevents onlookers from entering the market to help the industry blossom.

These activities are somewhat despicable because they give a whole new meaning to the definition of organized crimes. Days are coming when the appropriate regulatory guidelines will be put in place to eliminate crypto market manipulation conducted via social media channels, which are currently carried out openly.

At this point, two points should hopefully be clear to the readers:

- How ICOs are launched
- ICOs are risky investments and should be handled with extreme caution

For now, regulators are focusing on vigorously policy the ICO market itself, and not so much the pumps and dumps channels which are by side results as opposed to an integral part of the ICO process. Pumps and Dumps are also easier to perpetrate on newer and hence smaller caps coins, originating from ICOs, that have not proven they can stand the test of time.

Some notably successful ICOs' fundraising (IcoDrops 2019)

- Ethereum: Raised USD16 million in August 2014
- Stellar: Raised USD39 million in August 2014
- EOS: Raised USD197 million in June 2017
- Tezos: Raised USD228 million in July 2017
- Telegram: Raised USD1.7 billion in February 2018

4 A New Age of Financing, Birth of the STOs

In 2018, the SEC[14] issued a guideline (constantly being updated) mentioning that all Coin Offerings are security tokens. Many regulators followed suit shortly after, which led to the development of a new form of Coin Offerings dubbed the STOs or Security Token Offerings.

Before we deep dive into an understanding of STOs and how they differ from ICOs, we must understand the classification of various coins in the world of cryptocurrencies.

4.1 Coin Classification Framework

The Cryptocurrency Like bitcoin and ETC has one and only one purpose, replace fiat currency (money) and enable the physical or digital exchange of value, digitally. With a cryptocurrency, I can buy a good, a service from you and vice-versa.

The Utility Token Like the tokens used in amusement parks to access rides, utility tokens serve one and one purpose only, to give you access to a blockchain platform. *The Basic Attention Token (BAT)* discussed earlier is one of them. Its purpose isn't to help you by goods or services. Its purpose is to enable the use of the BAT dApp.

The Security Token The agenda of the security token is very much investment driven. Its purpose is to give its holder, some or all of the same rights that an investment security confers. *SpiceVC*, (Crunchbase 2019), for example is a security token that entitles the holder to share in the exit revenues of an existing Venture Capital firm's portfolio of private investments. In Feb 2018, SpiceVC raised USD20 million in funding from accredited[15] investors.

Breaking it down further, a security token might or might not have any or all the following characteristics:

- *Equity token*: by holding this token, you own a share of the company behind the project or a stake in the project itself. The share of the profit, company or project you receive depends on the structure of the token offering. You might also be

[14]In its published framework on Digital Assets i.e. cryptocurrencies and tokens, the US Securities Exchange Commission (SEC), which regulates securities markets, outlines that all coins (at the exception of Bitcoin and Ethereum) are securities until proven otherwise. Therefore, ICOs should by default fall under the umbrella of securities' fund raising and must meet the appropriate regulatory guidelines or risk of suffering severe legal consequences. https://www.sec.gov/corpfin/framework-investment-contract-analysis-digital-assets.

[15]Accredited investors (can be individual or institution), as per the SEC's definition, should have a minimum of $200,000 income in the past 2 years and expected in the current year. Alternatively, an accredited investor should own $1 million in net worth excluding the primary home of residence.

entitled to voting rights depending on what is stated in the whitepaper. These concepts are very similar to the traditional Equity market.

- *Debt token*: by holding this token you are entitled to an interest portion on the amount lent to the company or to the users of the asset owned. Depending on the structure and purpose of the project, you might expect the principal to be returned at a certain date, or not. There are some similarities with the traditional Fixed Income market. *BlockMason*, (Blockmason 2019), with its Credit Protocol system is one such debt token enabling a decentralized peer-to-peer lending and expense sharing system. According to *ICOHolder*, it raised close to USD 1.6 million in 2017.
- *The Asset Backed token*: this token gives you the right to own a specific asset and share in the revenues it generates. This is a security token that is representing an existing physical asset that might be otherwise illiquid such as a diamond mine, a piece of real estate, art collectibles, intellectual property, commodities such as sugar or oil and even a currency. In other words, any non-digital and physical asset is digitized via tokenization and rendered liquid thanks to its tokens. For example, *Digix* is an asset backed token that tokenized physical gold, (Digix 2019). The gold is held in vaults, and 1 DGX token = 1 Gram of Gold. Digix was launched on the Ethereum blockchain in 2014 and raised USD5 million.
- *The StableCoin*: in its most basic form, a stablecoin aims to bring stability back to an historically volatile cryptomarket. Critics of the original bitcoin have claimed that Satoshi failed to achieve her vision of replacing the US dollar with a currency backed by the faith of the people. They argue that due to the volatility of the bitcoin and other cryptocurrencies that followed suit, the people are not able to use cryptocurrencies to buy daily staples such as a loaf of bread or a gallon of milk.

To counter these arguments, the blockchain industry rose to the challenge and created the "*StableCoin*", a coin that does not fluctuate. For instance, the *Tether* (USDT), the stablecoin with the largest market cap (USD3 Billion) and hence the most widely used, is pegged one on one to the USD. The creators of Tether launched it on two blockchain platforms, Omni and Ethereum, (Tether 2019). At the time of writing, it was reported that they are planning to also launch on EOS. Tether also has a Euro backed stablecoin (EURT).

There are various forms of StableCoins which we will not explore in details in this chapter, but suffice to say that a stable coin is a form of asset backed token that comes with guarantees. Some guarantees such as: for each token held, one is entitled to 1 USD, or 1 CAD. Some stable coins take it further by applying the concept of *investment composite* and *indexes* to their structure. Meaning, for each one token held, a holder is entitled to a basket of cryptocurrencies. You see the point.

One stablecoin that caused a lot of ink to flow is the *Gemini dollar* (GUSD), (Gemini 2019), a token created by the Winklevoss twin brothers on the Ethereum blockchain platform. The GUSD is pegged to the US dollar and guarantees that 1 GUSD will always equal to 1 USD regardless of the crypto market fluctuation. You certainly wonder what the logic behind using a USD token as opposed to the USD dollar itself might be.

Proponents of currency backed stable coins argue that, USD in its tokenized form, helps merchants leverage the reduced transaction fees and freedom of international movement provided by decentralized solutions.

They maintain that using centralized forms of payments perpetrate the fees of the various middlemen such as VISA and Mastercard, working behind the scenes, and needed to settle each and every transactions.

Critics from the decentralized blockchain industry argue however that using a stablecoin, centralizes the decentralized since the actual USD must be held by a central authority, which in this case is the Gemini Trust Company LLC. and its partnering custodian State Street.

In practice however, the value of a stablecoin is in the eyes of the beholder. Depending on the intended purpose, some might find this stablecoin vs. another a better choice, and some might simply want to stick to actual USD if more convenient.

Readers must recall that a security token just like any other token, might give one the right, but not the company's legal obligation to uphold any end of the bargain outlined in the whitepaper. The Hybrid Token combines any of the above token characteristics. A hybrid token could be a cryptocurrency and a utility token. For example, Ethereum can be used as digital money, or used to create dApps on the Ethereum blockchain. A hybrid token could be utility token and a security token (equity token). For example, Peerplays (PPY) token, which gives you the non-legally enforceable right to share in the revenues of the platform, while using the platform to place gaming bets.

We should note here that a security token can also be structured to mimic the workings of derivatives.

4.2 *The Dawn of the STOs*

Amid the crypto winter, a few projects that tried launching ICOs, miserably failed as investors recovering from their initial excitement in ICOs and subsequent 2018 losses withdrew from the crypto market. This situation combined with the SEC declaring that all ICOs are Security tokens unless proven otherwise, led to a change in behaviours from aspiring ICO start-ups.

4.3 *Avoiding the Mistakes of the Past*

In 2019, for most legitimate projects looking to raise funds using tokens, there is now a strong focus on avoiding the mistakes of the past. The more a decentralized project includes some of the following non-exhaustive criteria, the more chance it will have to capture investors' attention:

(a) Reducing the uncertainties surrounding ICOs by presenting, in addition to the whitepaper, a tested POC/MVP (Minimum Viable Product)
(b) In addition to the caliber of the team behind the vision, showing a clear path to revenue instead of making the fund raising the end and ultimate goal of the project
(c) Giving out equity! Gone are the days when Jane and Joe could simply raise funds from investors without giving out some legally enforceable equity
(d) As a result of the above points, increasing security regulatory compliance

The shift in expectations from both investors and regulators led to the launch of the Security Token Offering. A step-up from the Initial Coin Offering method of fundraising. By definition, the STO is a regulated security restricted to accredited[15] investors[10] (Dan Handford 2018). Essentially, it is an exempt security, filed with the regulators of the chosen jurisdiction that happens to be exercised using a permissioned or private blockchain system.

We should mention that a security token does not need to be issued via an STO. A security token by design can and prior to 2018 had been issued without any regulatory approval or fillings. In fact, many tokens which were securities by design were launched via ICOs_security token by design does not mean legal and regulatory compliance. The process of issuing an STO today on the other hand must strictly follow and meet regulatory requirements.

For projects, looking to launch STOs (Fintech4Good 2019) that are U.S. compliant, there are several regulatory frameworks to follow depending on the structure of the STO as well as the type of investors the project wishes to access. Non-US jurisdiction such as Malta, Singapore and Switzerland follow their own frameworks. Projects incorporated in the U.S. or looking to raise from U.S. investors should follow one of the following regulations: Reg A+, Reg CF, Reg D or Reg S. Most popular Security Tokens Offerings were issued under Reg D and let's briefly review its implications.

Regulation D

Under Reg D, a company must file Form D with the SEC but securities' registration is not required, (SEC 2017).

For example, note the following statement in the SpiceVC (2017) investment memorandum: "The Spice Tokens have not been and will not be registered under the Securities Act of 1933, as amended (the Securities Act), or any other law or regulations governing the offering, sale or exchange of securities in the United States or any other jurisdiction."

Under Reg D, (Investor.gov 2019) the project can raise $5 million and above with restrictions. For $5 million and less the investor does not need to be accredited[15], but the number of non-accredited investors involved is limited. Non-accredited investor should be sophisticated[16] (James Chen 2019). Anything above $5 million, should involve accredited[15] investors only. Securities are subject to a resale lockup period. The project under Reg D might be subject to and must comply with state regulations however more stringent than federal regulations.

Readers note that the purpose of this chapter is not to provide investment or legal advice, but rather to give insightful information on the workings of alternative sources of financing in the context of the blockchain ecosystem.

Some successful STOs registered under Reg D include:

- SpiceVC, as already mentioned, tokenized the VC investment portfolio, and raised $15.5 million.
- Aspen Coin tokenized the St. Regis Aspen Resort in Colorado, by offering stakes in property. It raised $18 million.
- Art Token tokenized artwork allowing multiple investors to hold a specific share of value in the object. The project raised $5.5 million.
- Smart Valor, a Switzerland based company looking to democratizing the access to wealth by tokenizing a basket of assets. Smart Valor raised 1.5 million in Swiss Franc.
- Braid Token tokenized the feature film of the same name BRAID and raised $1.5 million.
- To date one of the biggest STOs came from tZero with $134 million raised to provide investors with fully executable SAFE agreements.

4.4 Steps for Launching an STO

The steps into launching an STO, (Newtown Partners Inc. 2019; Fintech4Good 2019), hold some similarities with the steps involved in launching an ICO, minus the regulatory requirements of an STO.

We can include the following:

1. *Conceive project*: Outline vision with clear path to revenue post fund raising. Prepare a minimum of a POC/MVP by engaging the team and partners behind the vision, expert consultants and tech companies.
2. *Structure the offering*: Assemble a team of legal experts, management consultants and advisors, broker-dealer, underwriter to structure the deal and file appropriate forms with the appropriate regulators. Receive regulatory approvals where applicable.

[16]A sophisticated investor is a high-networth investor who is considered to have a depth of experience and market knowledge in business matters to evaluate the risks and merits of an investment. This makes them eligible for certain benefits and opportunities.

3. *Market the project*: Create Investors-tailored marketing materials, including investment memorandums.
4. *Build STO Technical Platform*: Choose appropriate blockchain platform, build smart contracts with relevant compliance guidelines and triggers, lockup and burn rules, token recovery procedures, and other necessary technical requirements.

   This specific step, I must add, is the reason why a company might prefer launching an STO as opposed to raising funds, through traditional methods under Reg D. The substantial cost that is removed, and security that comes with automating the rules and compliance of fundraising is a very appealing proposition.
5. Launch Roadshow to *meet and pitch to investors*.

Post-fundraising, STOs must also strive to list their tokens on an exchange, for liquidity purposes. This is not the easiest enterprise given the limited number of exchanges currently built to onboard STOs. However, given the resale lockup period involved in STOs, exchange listing shouldn't be immediate cause for concerns. This is also not a concern, if the STO is launched on a private or public exchange.

4.5 Advantages and Disadvantages of STOs vs. ICOs

Right off the bat, the advantages of running an STO vs. an ICO are evident.

To begin with, STO investments do not rely on snake oil sales. The project must prove that the vision is viable by providing the data, prototype, POC/MVP and other elements to back it up. Especially for projects looking to bring liquidity to a physical asset via tokenization, there is palpable proof and better understanding of the reasons behind the fundraising. In this case, investors can demand to see legal proof of the physical asset's ownership before engaging in any further discussion. Contrary to the ICO market where 2/3 of the projects turned out to be scams, in the STO market there is a higher expected success rate post-offering.

In addition, the mandatory regulatory and legally enforceable requirements imposed on STOs mean that individuals with fraudulent objectives in mind are more likely to be kept at bay. Overall, this is good news for a crypto market that has been desperately pushing for mainstream adoption.

The lower risk associated with STOs could open the floodgate to institutional investors money, especially investments from banks. Banks are some of the most heavily regulated institutions in the investment industry due to their direct impact on entire economies. Having watched the crypto market from afar, STOs could be their entry point into what might become the prevailing fundraising method of the future.

Why Aren't There More STOs?

There are several reasons that slow down the use and adoption of STOs when compared to rate experienced during the golden age of ICOs in 2017. Some of these reasons are:

(a) The lack of understanding of STOs and their implications means that the blockchain ecosystem and onlookers are still threading with caution.
(b) It has taken the whole of 2018 to establish a certain regulatory clarity with the crypto innovative approach to investing and fundraising, and in some jurisdictions, clarity is still being defined.
(c) Launching an STO is equivalent to using traditional fundraising methods, meaning unlike the ICO market, not all traditional fundraising fees are removed from the equation.
(d) STOs are restricted for the most part, to accredited[15] investors, cutting out the global public market, which has called for criticism from proponents of the purest form of a decentralized marketplace.

It is my opinion that, in time, the STOs and ICOs markets will cohabit, as complementary methods of fundraising.

The SEC has recommended the use of the Howey test (FindLaw' Team 2019) to determine whether a crypto project will potentially be considered a security token or a utility token, with the former subject to registration and/or regulatory fillings. If the project meets the Howey Test's definition of an investment contract, then it is a security token. The Howey test (Olivier Dale 2018) states that if a project qualifies as (1) an investment of money, (2) with an expectation of profit, (3) in a common enterprise, (4) with the profit being generated by a third party, then it is an investment contract.

I believe that the future of cryptos will be divided into two fundraising steps:

- Phase 1: Launch an STO to raise funds that will be used to build a fully flushed out and functional blockchain platform. Reward accredited[15] and sophisticated[16] investors, for early investment and taking on substantial risk by allocating these investors with legally enforceable equity, stakes, ownership of the project or company.
- Phase 2: Launch an ICO providing public, non-accredited investors turned users, access to the ready platform, with ideally investment caps. Funds shall be used to maintain and provide ongoing development and upgrades to the Blockchain or non-Blockchain platform.

Not all projects will need to go through all 2 phases. For example, some Asset Backed Security token might end at Phase 1, while some utility token project might only undertake Phase 2. Under the assumptions that a project will not need to launch an STO if the team could self-fund the fully functional blockchain platform prior to the ICO. In fact, an existing company, with a proven record of revenues and a significant traction and a working fully functional product, might be a great candidate for an ICO, which could be used to accelerate its expansion.

5 IEO, Innovating in an Emerging Industry

The Initial Exchange Offering (IEO) is a 2019 innovative approach to raising funds similar but unlike ICOs.

5.1 *The IEO Process*

As the name implies it, exchange offerings are token fundraisings exclusively conducted on a crypto exchange (Benjamin Vitaris 2019). In other words, a start-up looking to raise funds, will approach a crypto exchange such as Binance, Bitfinex, OkEx and others already previously mentioned. The exchange will perform their rigorous due diligence on the team and projects. Once approved, the exchange handles the ICO process from beginning to end, with the exception that the project is exclusively marketed to its user base. For an investor to have a chance to access the project's tokens, the investor will need to open an account on the exchange.

Essentially, the exchange is acting as the fundraising underwriter, while providing a readily available market of token buyers to the start-up or project.

This method of fundraising is a *win-win-win* for all parties involved.

- *A win for the start-up* who once approved has its pre-fundraising efforts including technical requirements, smart-contract construction, marketing and investor materials and KYC handled by a one-stop-shop, reliable crypto exchange. Post-fundraising, the start-up needs not worry about listing on an exchange as the tokens are automatically listed on the very same exchange that led the raise.
- *A win for the investors*, as they can trust that the exchange will and has performed the necessary due diligence on the solidity and reliability of the start-up team or founders prior to onboarding the project. They can also readily trade the tokens as opposed to waiting months for exchange listings.
- *A win for the exchange*, as the exchange charges listing fees and takes a percentage of the funds raised, incurring an additional source of revenue. By the same token (pun intended) , the exchange gains new users, as outside investors wanting access to the project, will create new accounts on the exchange. Therefore, every time a new IEO is launched, the exchange might see its user base substantially increased which in turn plays as a great bargaining chip on the next IEO. Finally, if the exchange has its own token, it might see its price appreciate as a result of each successful IEO launched on the platform.

IEO Launchpads

Most exchanges jumping on the IEO waggon, create *launchpads* dedicated entirely to bringing an IEO from beginning to end. Some notable launchpads include the *Bittrex International IEO*, the *Huobi Prime* and the *OKEx IEO*. The very first IEO launched Feb 2019, was the *BitTorrent* (BitTorent 2019) token sale which used the Binance Launchpad (Binance is a top crypto exchange). In less than 15 min, the fundraising had met its hardcap (max funding goals) of USD7.2 million. BitTorrent is an existing, previously non-tokenized platform, that allows content creators to connect with their users. With the tokenization of the platform, users will be able to exchange tokens for a faster download experience amongst other benefits. Other notable successful 2019 IEOs include the *Newton Project*, which raised USD28 million on *Huobi Prime*, and *Blockcloud* which raised USD2.5 million on OKEx IEO. Despite their apparent success post-crypto winter, IEOs are not without their fair share of cons.

Contra-arguments Against IEOs

While it is a very convenient way of raising funds, process wise, *cost wise* it is not clear whether IEOs are a cheaper alternative to ICOs. Launchpad listing fees are expensive and can go as high as 20BTC (USD173,800 as of June 1, 2019 on CCMC). In addition, the percentage cut of proceeds from the fundraise can reach as high as 10% which when compared to the $7.2 million raised by BitTorent come close to USD800,000. This amount is not far from the cost of running an ICO these days, minus the down-payment. Factoring in the convenience of the IEO process, running an IEO might still be a great proposition for many start-ups.

Furthermore, it is not clear what *auditing process* are put in place to avoid a potential conflict of interest between the exchange and the listing company. For a legitimate exchange, we would like to think preserving reputation would however be on top of the priority list. On that note, Binance cancelled the RAID's IEO hours prior to launch time as a flaw was found in the business model. We would hope that going forward all exchanges launching an IEO will adopt the same principles, always putting investors first.

In addition, centralized exchanges are subject to cyber attacks as it has been the case in the past with Binance. Therefore, for an IEO launching exclusively on an exchange, the security of funds collected should be a concern. The more reputable and largely used the exchange, the more attractive to hackers, and the more preventive, high grade security measures become of utmost importance.

As for investors, with the instant access to an exchange where they can sell their tokens, measures should be taken to avoid dumps that depress the price of a token. So far, many of the IEOs have experienced constant growth in token price, post-raise, which might be attributed to many factors beyond the scope of this reading.

Finally, given that IEOs are new, it is not clear where regulators stand on the topic, and we might expect future clarifications and/or restrictions imposed on this innovative method of financing.

6 Conclusion

In conclusion, the blockchain technology and associated crypto market have created new and alternative sources of financing that cannot be ignored. While the industry is still figuring itself out and innovating, not so old methods such as ICOs and STOs are still being defined, while newer methods such as IEOs are emerging. As the blockchain industry matures, the future shall hold a colourful array of reaped financing and investment avenues, bringing flexibility and affordability to the marketplace. Ultimately, these innovative approaches to financing will benefit investors, companies, institutions and consumers, or in other words the global economy as a whole.

References

Arcona (2019) https://www.arcona.io/
AxiomZen (2019) https://www.axiomzen.co/about
Basic Attention Token (BAT) (2019) https://basicattentiontoken.org/
Bitcoin (2019) https://bitcoin.org
BitTorrent (2019) https://www.bittorrent.com/company/about
Blockmason (2019) https://blockmason.io/
Bloxy (2019) https://bloxy.info
Chen J (2019) Sophisticated investor. https://www.investopedia.com/terms/s/sophisticatedinvestor.asp
Clay M (2018) Blogpost, whitepaper versus yellowpaper: what is the difference? https://cryptocanucks.com/whitepaper-versus-yellowpaper-what-is-the-difference/
Corporate Finance Institute Resources (2019) IPO process. https://corporatefinanceinstitute.com/resources/knowledge/finance/ipo-process/
Crunchbase (2019) SpiceVC section funding rounds. https://www.crunchbase.com/organization/spice-vc#section-funding-rounds
Cryptokitties (2019) https://www.cryptokitties.co/
Dale O (2018) What is the Howey test & how does it relate to ICOs & cryptocurrency? https://blockonomi.com/howey-test/
Digix (2019) https://digix.global
District0x Educational Portal (2019) ERC721 tokens (non-fungible tokens) explained. https://education.district0x.io/general-topics/understanding-ethereum/erc-721-tokens/
Ethereum Foundation (2019) https://Ethereum.org
FindLaw' Team (2019) https://consumer.findlaw.com/securities-law/what-is-the-howey-test.html
Fintechforgood (2019) https://fintechforgood.co
Gemini (2019) https://gemini.com/dollar
Giel D (2019) CNBC, Federal judge tells traders they can combine cases accusing JP Morgan of rigging metals market. https://www.cnbc.com/2019/02/07/federal-judge-tells-traders-they-can-combine-cases-accusing-jp-morgan-of-rigging-metals-market.html
Green L (2019) Fungibility. https://www.investopedia.com/terms/f/fungibility.asp
Handford D (2018) Accredited investor vs. sophisticated investor. https://www.biggerpockets.com/member-blogs/10864/76594-accredited-investor-vs-sophisticated-investor
IcoDrops (2019) https://icodrops.com/ethereum/
Mitra R (2018) What is tokenomics? Ultimate investor's guide - part 1. https://blockgeeks.com/guides/what-is-tokenomics/
Newtown Partners Inc. (2019) Security token primer. https://newtownpartners.com
Noor (2019) MeetNoor, Ethereum Dapps list. https://meetnoor.com/ethereum-dapps-list/
SEC (2017) Fast answers, Reg D. https://www.sec.gov/fast-answers/answers-regdhtm.html
SpiceVC Venture Capital Pte. Ltd., Investment Memorandum (2017) https://docsend.com/view/42v5w5c
Tether (2019) https://tether.to/
U.S. Securities and Exchange Commission (2019) Rule 506 of regulation D. https://www.investor.gov/additional-resources/general-resources/glossary/rule-506-regulation-d
uPort (2018) https://www.uport.me/

Vitaris B (2019) BlogPost, What is an initial exchange offering (IEO) and how it differs from ICO? https://cryptopotato.com/what-is-an-initial-exchange-offering-ieo-and-how-it-differs-from-ico/
WikiCryptoCoins (2018) Yellow paper. https://wikicryptocoins.com/currency/Yellow_Paper
Wikipedia (2019) White paper, the free encyclopedia. https://en.wikipedia.org/wiki/White_paper
Wikipedia, ERC-20 (2019) The free encyclopedia. https://en.wikipedia.org/wiki/ERC-20
Yu J (2018) Differences between a white paper, yellow paper, and beige paper. https://medium.com/@hello_38248/differences-between-a-white-paper-yellow-paper-and-beige-paper-ad173f982237

Tokenomics

Ralf Wandmacher

Abstract The tokenomics of Initial Coin Offerings is a new field of research. The wording Initial Coin Offering is only a few years old, the field of tokenomics is even younger. This chapter will discuss the parameters of the tokenomics of Initial Coin Offerings in a qualitative and quantitative way to gather knowledge about upcoming standards and the definition of current requirements. As more and more Initial Coin Offerings coming to the market, a fundamental view of tokenomics is required. This chapter identifies 13 important parameters of tokenomics. Each of them is examined by literature and the used sample data set. Further parameters are identified as well as further research objectives in the field of tokenomics.

1 Introduction

Initial Coin Offerings raise funding through the creation of token by smart contracts. The newly created token (or minted token) are partly sold to fund ideas and networks. A set of parameters is used to define these new tokens to create an economic value. This definition can be called tokenomics. Hence, tokenomics describe the function and parameters of the offered token in an Initial Coin Offering (ICO) process. The tokenomics is a fundamental part of each ICO. In the tokenomics of ICOs the economic design of the offered token is defined by applied parameters.

The aim of this chapter is to identify important parameters. The research is based on two steps. First, the most important parameters in tokenomics will be identified by a qualitative discussion of literature. Second, the identified parameters will be examined quantitatively on a set of selected ICOs. The aim of the research is to identify the most important parameters of ICOs and to quantify these parameters as usable parameters for future ICOs.

The used set of data of ICOs consists of 98 closed and announced ICOs. It is only a small part of the overall ICO universe and the data is skewed with 88 data sets to 2018. The set of data is not fully representative, but may give hints how the selected

R. Wandmacher (✉)
Accadis University of Applied Sciences, Du-Pont-Str. 4, Bad Homburg, Hesse, Germany
e-mail: ralf.wandmacher@edu.accadis.com

S. Goutte et al. (eds.), *Cryptofinance and Mechanisms of Exchange*, Contributions to Management Science, https://doi.org/10.1007/978-3-030-30738-7_7

qualitative parameters have been used during 2018. Further research may apply to this selection of qualitative parameters to gain further insights. The data was collected from the respective whitepapers of the projects and from sources as icobench.com, icorating.com, icodata.io and tokenmarket.com.

2 Initial Coin Offerings

ICOs do have a main advantage for the founders: ICOs do not dilute their equity stake for financing (Kaal 2018, p. 2). Other advantages are freeing investors from their home bias, and removing the need for financial intermediaries (Boreiko and Sahdev 2018, p. 3).

Also the high cost of traditional Initial Public Offerings (IPOs) with 4–28 million USD in the US (Preston 2018, p. 328) directs the way to ICOs. The costs of ICOs can be ten times lower than IPOs (Dell'Erba 2017, p. 11) or ICOs are even seen as a low-cost fundraising option (Lipusch 2018, p. 11). Whereas the IPOs are often seen as exit strategies, the ICOs are more entry strategies to finance the idea (Felix 2018, p. 5). ICOs, acting as a new form of investment, create "significant level of information asymmetry" (Felix 2018, p. 10). In particular, the information asymmetry in ICOs is pronounced for small investors (Fisch 2018, p. 6).

Long (2018) states that ICOs offer "low friction costs—there are no underwriters, trustees, transfer agents, exchanges, custodians, clearinghouses or central securities depositories". ICOs disrupt the traditional capital market in the venture capital and investment bank world. ICOs raised 7.2 billion USD in the second quarter of 2018, 45% of the amount of traditional Initial Public Offerings (IPOs) and 31% of traditional venture capital in the second quarter of 2018 (Long 2018).

3 History of ICOs

The phrase Initial Coin Offering and ICO was created by the US-based gaming start-up "Breakout" in November 2014 (Boreiko and Sahdev 2018, p. 13). The ICO history started even before 2014 with 2 ICOs raising 630 k USD in 2013. In 2014, 11 ICOs raised 33 million USD, 14 ICOs raised 11 million USD in 2015 (Boreiko and Sahdev 2018, p. 11). In 2016, 74 ICOs were realised and in 2017 more than 1000 ICOs (Benedetti and Kostovetsky 2018, p. 9). However, 37% of all proceeds were made by only 20 ICOs (Momtaz 2018a, b, p. 9).

The crowdfunding company Kickstarter raised 3.6 billion USD since its inception in 2009. In contrast, ICOs raised 7.5 billion USD in 2017 alone (Amsden and Schweizer 2018, pp. 2–3). In total, more than 18 billion USD were raised through ICOs between 2014 and June 2018 (Howell et al. 2018, p. 1). Momtaz (2018b) even states a number of 21 billion USD in ICO fund raising from 2013 to 2018. The success of ICOs is disrupting the traditional centralized venture capital and investment bank models (Boreiko and Sahdev 2018, p. 10).

4 Literature Review of Parameter of Tokenomics

The parameters of tokenomics are important for ICOs. In the literature different parameters are discussed and even some parameters found in white papers have not been discussed in the reviewed literature so far.

One of the main discussion is about the token type, i.e. is it an utility, security, or payment token (or medium of exchange and store of value "coins") according to Howell et al. (2018, p. 1). The discussion about the segmentation is extensive, the Swiss regulator FINMA classifies three token categories—payment tokens, utility tokens, and asset tokens (FINMA 2018, p. 3). Howell et al. (2018, p. 4) find that 68% of all token in their sample are utility token. The token type is identified as first parameter.

A further parameter belongs to the used technical standard of the token. The technical standard depends on the use of the respective blockchain. For example, EOS, NEO, Ethereum are blockchains from which tokens could be used in the ICO.

Momtaz (2018a, p. 8) finds a market share of more than 80% for ERC20 (Ethereum Request for Comment 20) tokens. Other research shows that 74% of tokens using an ERC20 definition (Howell et al. 2018, p. 23). Felix (2018, p. 20) finds in the examined data set even a use rate of 84% of the Ethereum platform. Adhami et al. (2018, p. 13) find only an Ethereum use rate of 56.5%.

Overall, the majority of all tokens use the ERC20 standard. The technical standard is defined as the second parameter.

Another important parameter is the standard issue price at a public ICO. A median value of 0.30 USD was found to gather behavioral investors attracted to low nominal prices (Benedetti and Kostovetsky 2018, p. 16). Felix (2018, p. 22) finds as issue price a median value of 0.20 USD. The public ICO price is also used as reference price for private sales. The standard issue price is set as third parameter.

An open question is in which currency the ICO is prized. The main differentiation is between fiat and crypto currencies. If the segmentation between these two is set, the open question is in which currency of the segments (i.e cryptocurrency or fiat) the ICO is priced. The data set sample will be examined for the standard currency, being the fourth parameter.

The question about the standard currency leads to the question which currencies are accepted as means of payment for the ICO. The acceptance of fiat currencies can be seen positively (connections to the traditional banking system) and negatively (lack of confidence to complete the ICO with cryptocurrencies, no protection of a soft-cap through smart contracts), overall the acceptance of fiat currency increases the uncertainty of the venture (Amsden and Schweizer 2018, p. 19). Momtaz (2018a, p. 17) finds that ICOs accepting fiat raise on average more as it reduces the barrier to entry. However, accepting Bitcoin (BTC) or Ether (ETH) has a positive association with success, 66% accept Ether but only 10% USD (Howell et al. 2018, p. 25). The literature review does not show a definitive result for this important parameter, currency acceptance, which is selected as the fifth parameter.

ICOs were done initially without regulatory processes and any reference to processes as Know Your Client (KYC), but starting in 2017 a growing number of ICOs were using KYC processes (Smith + Crown 2017). Tezos even demanded a KYC 11 months after the ICO ended (Devoe 2018). As KYCs are done regularly today, the cost for it require minimum contributions to the ICO. Minimum contribution requirements signal that the founders are confident in the quality of their offering (Amsden and Schweizer 2018, p. 20). The minimum contribution is applied to the public ICO sale.

Before this public sales phase starts, the private sale may sometimes be executed. The private sale is only open to accredited and institutional investors who have to invest high minimum contributions (e.g. 100,000 EUR at Helix Orange (2018)). Private sales are used to attract sophisticated investors (Amsden and Schweizer 2018, p. 37) and to cover the setup costs of ICOs (Amsden and Schweizer 2018, p. 17). 44% of ICOs conduct a private sale (i.e. pre-ICO sale) according to Momtaz (2018a, p. 8) and 36% according to Felix (2018, p. 22). Benedetti and Kostovetsky (2018, p. 16) find that 40% of ICOs hold a private sale. A private sale and the pre-ICO equity investment of venture capital companies (VCs) may signal quality (Howell et al. 2018, p. 3). In their sample, 45% of ICOs use a private sale to fund the ICO, certify the issuer and to determine demand (Howell et al. 2018, p. 12). Quite interestingly, Felix (2018, p. 10) finds that private sales reduce underpricing.

The literature research shows clearly the importance of the parameter private sale to be selected as the sixth parameter. The selling phases are often incentivised by bonus or discounts. The earlier the investor commits the capital to the sale of the token, the higher the bonus or discount. A bonus or discount in the private sale can be related to the success of the ICO, but no evidence is found in the literature (Amsden and Schweizer 2018, p. 37). Adhami et al. (2018, p. 13) finds a bonus in 54.8% of the examined ICOs.

The sales process furthermore consists of the observable sales duration of the public ICO. Private sale phases are often not stated, hence they are less observable. Benedetti and Kostovetsky (2018, p. 16) find that the average ICO takes 37 days with a median of 31 days, however the average was rising to 41 days for ICOs in 2018. The public sales duration is identified as seventh parameter.

The sale process requires also the overall amount of token to be sold. The overall amount consists of the sellable token plus the other distributed tokens plus the in future minted tokens. Bitcoin will getting close to a total supply of 21 million token, XRP will have a maximum supply of 100 billion token. This eighth parameter is the total supply.

Some ICOs do use minimum funding amounts, so called soft caps, and maximum funding amounts, so called hard caps. The soft cap is seen as a protection mechanism for investors, i.e. if the amount of the soft cap is not reached all contributions are returned to the investors. This reduces uncertainty and decreases investor risk (Amsden and Schweizer 2018, p. 19). Research shows that a hard cap increases

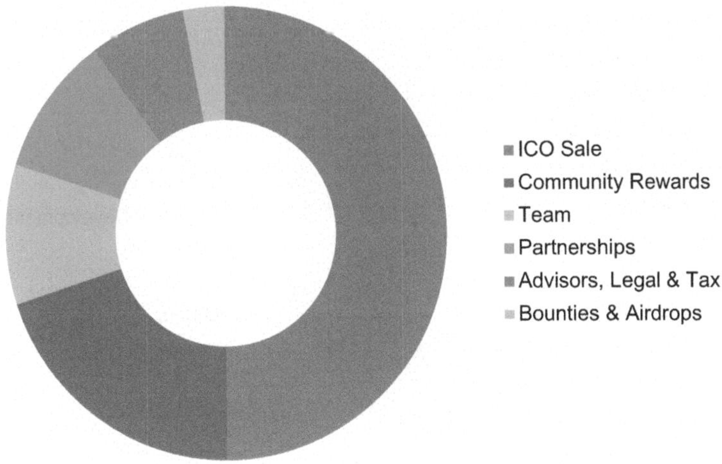

Fig. 1 Token allocation

the probability of token tradability and the amount raised in the ICO (Amsden and Schweizer 2018, p. 37). The ninth parameter is the soft and hard cap.

The sale process requires also the quota of tokens to be sold in relation to minted tokens. The sellable token amount is set in relation to the overall number of token to get the total supply number in percentage terms. Benedetti and Kostovetsky (2018, p. 16) find that 60% of all tokens are sold on average during the ICO. Howell et al. (2018, p. 25) find that 54% of total token supply is sold. Conley proposes to sell all tokens available and use parts of the proceeds for business development, salary payment and other uses instead of holding back tokens (Conley 2017, p. 10). The sale quota may show the investors if their purchased tokens will be diluted in future, hence the proposal of Conley can be seen as a try to prevent future dilution of the purchased amount of tokens. The sale quota is identified as tenth parameter.

Not all the tokens are sold, but allocated to different parties. The token allocation shows where the overall token supply is allocated to. The token allocation is a main part of the white paper. Often, it is illustrated as a doughnut. Token allocation is defined as the eleventh parameter. An example of the token allocation can be seen in Fig. 1.

Investors are also interested in the distribution schedules of amount of token for the team (i.e. founders, team members) and the advisors. These distribution schedules are also termed vesting schedules. Howell et al. (2018, p. 23) describe that 36% of their examined ICOs have some kind of vesting schedule. Smith + Crown (2017) describe that vesting was rare up to 2016, starting in 2017 vesting schedules have a duration of up to 36–48 months.

Tokens allocated to the ICO are sold against fiat or cryptocurrency. Investors want to know where the money is going to in the project. The money may be used for marketing, technology, business development, legal, operations and other functions.

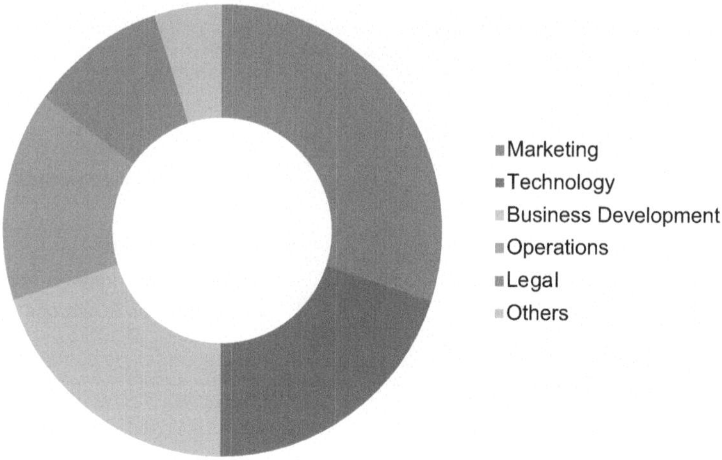

Fig. 2 Use of sale proceeds

This use of sale proceeds is the twelfth parameter and also often illustrated as doughnut, as shown for example in Fig. 2.

The creation of tokens is accompanied by the fact, that tokens are tradable. In contrast, venture capital investments are illiquid and it is difficult to monetize them in the near term, they may be monetized only through an exit (i.e. sale as through an IPO) after some years. In contrast, tokens could be traded immediately. Amsden and Schweizer find in their research that a tradable token is of the "utmost importance" (Amsden and Schweizer 2018, p. 14). However, it is not important for every ICO, e.g. Polkadot raised 140 million USD in Ether in October 2017 (Russo and Kharif 2017) with the announcement to release their "DOTs" in the third quarter of 2019 (Polkadot 2017).

Tradability also creates interest in ICOs so that capital can be collected. The overall amount of money collected in ICOs is large especially when compared to the raised amounts of money from Kickstarter in a much longer time frame. Also, founders and investors wants to understand how much money could be raised for their projects.

The average successful ICO raised 11.5 million USD with a median of only 3.8 million USD according to Benedetti and Kostovetsky (2018, p. 16). Howell et al. (2018, p. 25) found an average raise of 15.8 million USD in their data set. Hence, capital collected is identified as thirteenth parameter in this research.

In total 13 parameters were identified to play an important role in the field of tokenomics. However, each ICO is different, a clear standardization is not established yet. Further parameters may become important in individual ICOs or over time. Also, some of the defined and identified parameters may be not applicable to every ICO.

The identified parameters are:

1. Token Type
2. Technical Standard
3. Standard Issue Price
4. Standard Currency
5. Currency Acceptance
6. Private Sale
7. Public Sales Duration
8. Total Supply
9. Soft Cap and Hard Cap
10. Sale Quota
11. Token Allocation
12. Use of Sale Proceeds
13. Capital Collected

5 Research of Parameter of Tokenomics

The used set of data of ICOs was examined for the identified parameters. As the examined data set is skewed to 2018, it is interesting if there are differences to the data of the literature which is only partly from 2018.

Parameter 1: Token Type
There are 68% of utility token in the sample of Howell et al. (2018, p. 4) and 68.7% in the researched sample. In addition, there are 31.3% payment token in the examined sample and no other token types.

Parameter 2: Technical Standard
The literature shows a use rate of 56.5–84% of ERC20 tokens, the examined sample of this report finds even 87.7% of tokens using the ERC20-standard.

Parameter 3: Standard Issue Price
In the literature a median price of 0.20–0.30 USD per token was found, the data set shows a mean of 0.40 USD with a median of 0.10 USD in the sample.

Parameter 4: Standard Currency
The sample set exhibits that 76.54% of the ICOs are priced in USD, 19.75% are priced in ETH and 3.7% in EUR.

Parameter 5: Currency Acceptance
Accepting BTC or ETH has a positive association with success, 66% accept ether but only 10% USD (Howell et al. 2018, p. 25). The sample data shows that 34.44% of the ICOs only accept ETH, 20% accept ETH, BTC and other cryptocurrencies, and 16.67% accept ETH and BTC only. ICOs accept fiat only in 28.29% whereas ETH, BTC and fiat is 15.56%, ETH, BTC, fiat and other cryptocurrencies are 10%, and ETH and fiat is only in 3.33% accepted.

Parameter 6: Private Sales
The literature shows that in 36–45% of the ICOs a private sale is executed. The data set shows a private sale in 8.16% of the ICOs. However, as most whitepaper do not mention a private sales, and even if mentioned the conditions are often not stated to the public.

Parameter 7: Public Sales Duration
The data exhibits a median of 31.5 days as duration of the public sales which confirms the 31 days of duration found in the literature review.

Parameter 8: Total Supply
The median of total supply of token in the ICOs of the data set is 775 million token.

Parameter 9: Soft Cap and Hard Cap
The soft cap is in place in 66.3% of the ICOs, a hard cap in 84.7% of the ICOs. The relation between the soft and hard cap, i.e. the minimum amount raised to be viable and the maximum amount raised, is shown with a number of 17.3% for the soft cap relative to the hard cap or 5.78 times the soft cap to get to the hard cap.

Parameter 10: Sale Quota
The review of the literature exhibits a sale quota of 54–60%. The evaluation of the used data set finds a mean of 53.3%, quite close to be in line with the facts of the literature review.

Parameter 11: Token Allocation
The sample data set shows that 55.5% of the allocated tokens go to the ICO, 11.6% into reserves, 8.2% to the team, 8.1% to the business development and marketing, 3% to advisors, 2.9% to bounties and airdrops, and 11.7% of the allocation to other functions (e.g. referral programs, incentives, consultants, CSR, foundations, future release, charity and more).

Parameter 12: Use of Sale Proceeds
The development of the business is with 38% the leading target for the proceeds of the ICO sale. The marketing function follows with 27%, followed by operations with 14% of the proceeds. Finally, 6% go to legal and 15% to other areas (e.g. technology, reserves).

Parameter 13: Capital Collected
Mean values of 11.5–15.8 million USD and a median value of 3.8 million USD for the fund raising through an ICO were found in the literature review. In the used sample, a value of 5.97 million USD was found for the capital collected. The reasoning behind this difference could be the omission of huge ICOs as Telegram and EOS, which raised around 6 billion USD alone. Also, the increasing number of ICOs in 2018, may have decreased the overall amount per individual ICO.

An overview of the parameters and a comparison of the sample findings and the literature research findings can be observed on Table 1: Tokenomics Parameter.

Table 1 Tokenomics parameter

Tokenomics parameter	Sample findings	Research findings
1. Token type	68.7% utility, 31.3% payment	68% utility
2. Technical standard	94.9% ERC20	56.5–84% ERC20
3. Standard issue price	Median: 0.10 USD	Median: 0.20–0.30 USD
4. Standard currency	76.54% USD, 19.75% ETH, 3.7% EUR	–
5. Currency acceptance	34.44% ETH only, 20% ETH, BTC and other cryptos, 16.67% ETH and BTC, 28.29% fiat and crypto	66% ETH, 10% USD
6. Private sale	8.16% of ICOs	36–45% of ICOs
7. Public sales duration	Median: 31.5 days	Median: 31 days, mean: 37 days (41 days in 2018)
8. Total supply	Median: 775 million	–
9. Soft and hard cap	66.3% soft cap, 84.7% hard cap, soft cap is 17.3% of hard cap	–
10. Sale quota	Mean: 53.3%	54–60%
11. Token allocation	55.5% ICO, 11.6% reserves, 8.2% team, 8.1% business development and marketing, 3% advisors, 2.9% bounties and airdrops, 11.7% other	–
12. Use of sale proceeds	38% development, 27% marketing, 14% operations, 6% legal, 15% other (tech., reserves)	–
13. Capital collected	Mean: 5.97 million USD	Mean: 11.5–15.8 million USD Median: 3.8 million USD

6 Conclusion

This chapter shows that many parameters become important in the field of tokenomics. The literature review identified single parameters which are important to follow. Overall, a number of 13 parameters were used in the research of a data sample collected from white papers and resources as ICObench.com. The quantitative results of the 13 parameters were compared to the findings of the literature review. Some parameters as token type, public sales duration or the sale quota were confirmed.

Other parameters were different to the findings in the literature research, i.e. the higher use of the ERC20 technical standard, the lower median standard issue price, the accepted currency in ICOs or the lower amount of capital collected in the data set. Other parameters were introduced in the literature review and quantified in the research of the sample set. The standard pricing of ICOs in USD with 76.54%, the total supply with 775 million token or that 66.3% of the ICOs use a soft cap and even 84.7% use a hard cap. Also that the hard cap is higher by the factor of 5.78 to the soft cap. Furthermore, the token allocation and the use of the sale proceeds were quantitatively introduced through the examination of the data set.

By studying white papers it is becoming obvious that further parameter can be identified and should be researched. A proposal for further research is the following list:

- Bonus structures of ICOs
- Underwriting of ICOs in the secondary market
- Length to listing and its implication
- Length of the individual sale phases and the overall duration of the ICO sale
- The total token supply and its market effects
- Token supply in the ICO selling phases
- Vesting structures

Tokenomics is developing itself fast as a new field of finance research. This chapter shall help to uncover this new field area.

References

Adhami S, Giudici G, Martinazzi S (2018, May 9) Why do businesses go crypto? An empirical analysis of initial coin offerings. J Econ Bus. https://doi.org/10.1016/j.jeconbus.2018.04.001. Accessed 27 August 2018

Amsden R, Schweizer D (2018, April 16) Are blockchain crowdsales the new 'gold rush'? Success determinants of initial coin offerings (April 16, 2018). Available at SSRN: https://ssrn.com/abstract=3163849 or https://doi.org/10.2139/ssrn.3163849. Accessed 15 August 2018

Benedetti H, Kostovetsky L (2018, May 20) Digital tulips? Returns to investors in initial coin offerings. Available at SSRN: https://ssrn.com/abstract=3182169 or https://doi.org/10.2139/ssrn.3182169. Accessed 26 August 2018

Boreiko D, Sahdev NK (2018, July 6) To ICO or not to ICO – empirical analysis of initial coin offerings and token sales. Available at SSRN: https://ssrn.com/abstract=3209180 or https://doi.org/10.2139/ssrn.3209180. Accessed 26 August 2018

Conley JP (2017, June 6) Blockchain and the economics of crypto-tokens and initial coin offerings. In: Vanderbilt University Department of Economics Working Papers, VUECON-17-00008. http://www.accessecon.com/Pubs/VUECON/VUECON-17-00008.pdf. Accessed 24 Aug 2018

Dell'Erba M (2017, July 7) Initial coin offerings. A primer. The first response of regulatory authorities. Available at SSRN: https://ssrn.com/abstract=3063536 or https://doi.org/10.2139/ssrn.3063536. Accessed 27 August 2018

Devoe R (2018, June 14) Tezos demand investor KYC Info 11-months after end of ICO. In: Blockonomi. https://blockonomi.com/tezos-kyc/. Accessed 28 August 2018

Felix T (2018, June 1) Underpricing in the cryptocurrency world: evidence from initial coin offerings. Available at SSRN: https://ssrn.com/abstract=3202320 or https://doi.org/10.2139/ssrn.3202320. Accessed 25 August 2018

FINMA (2018, February 2) Guidelines for enquiries regarding the regulatory framework for initial coin offerings (ICOs). https://www.finma.ch/en/~/media/finma/dokumente/. . ./wegleitung-ico.pdf?la=en. Accessed 25 August 2018

Fisch C (2018, March 1) Initial coin offerings (ICOs) to finance new ventures: an exploratory study (March 1, 2018). Available at SSRN: https://ssrn.com/abstract=3147521 or https://doi.org/10.2139/ssrn.3147521. Accessed 24 August 2018

Helix Orange (2018, July 30) Whitepaper. Version 1.5 EN. https://ico.helix-orange.com/wp-content/uploads/2018/08/HELIX-Orange_Whitepaper_v1.5.pdf. Accessed 27 August 2018
Howell ST, Niessner M, Yermack D (2018, July) Initial coin offerings: financing growth with cryptocurrency token sales. In: Finance working paper no. 564/2018, European Corporate Governance Institute. http://www.ecgi.global/sites/default/files/working_papers/documents/finalhowellniessneryermack.pdf. Accessed 26 August 2018
Kaal WA (2018, February 2) Initial coin offerings: the top 25 jurisdictions and their comparative regulatory responses. CodeX Stanford Journal of Blockchain Law & Policy (2018); U of St. Thomas (Minnesota) Legal Studies Research Paper No. 18–07. Available at SSRN: https://ssrn.com/abstract=3117224 or https://doi.org/10.2139/ssrn.3117224. Accessed 26 August 2018
Lipusch N (2018, March 23) Initial coin offerings – a paradigm shift in funding disruptive innovation. Available at SSRN: https://ssrn.com/abstract=3148181 or https://doi.org/10.2139/ssrn.3148181. Accessed 25 August 2018
Long C (2018) ICOs were 45% Of IPOs in Q2 2018, as cryptos disrupt investment banks. In: Forbes. https://www.forbes.com/sites/caitlinlong/2018/07/22/icos-were-45-of-ipos-in-q2-2018-as-cryptos-disrupt-investment-banks/#5067f8f5794c. Accessed 25 August 2018
Momtaz PP (2018a, April 21) Initial coin offerings. Available at SSRN: https://ssrn.com/abstract=3166709 or https://doi.org/10.2139/ssrn.3166709. Accessed 22 August 2018
Momtaz PP (2018b, May 23) Putting numbers on the coins: the pricing and performance of initial coin offerings. Available at SSRN: https://ssrn.com/abstract=3169682 or https://doi.org/10.2139/ssrn.3169682. Accessed 29 August 2018
Polkadot (2017, September 20) Lightpaper. Version 1. https://polkadot.network/Polkadot-lightpaper.pdf. Accessed 22 August 2018
Preston J (2018) Initial coin offerings: innovation, democratization and the SEC. In: 16 Duke Law & Technology Review, pp 318–332. https://scholarship.law.duke.edu/dltr/vol16/iss1/10. Accessed 25 August 2018
Russo C, Kharif O (2017, December 12) The hottest ICOs are the ones that have done the least amount of work. https://www.bloomberg.com/news/articles/2017-12-12/want-to-issue-a-red-hot-ico-rule-no-1-is-do-very-little-work. Accessed 26 August 2018
Smith + Crown (2017, September 8) Trends in token sale proposals. https://www.smithandcrown.com/trends-token-sale-proposals/. Accessed 27 August 2018

Crypto Tokens and Token Offerings: An Introduction

Chen Liu and Haoquan Wang

Abstract This chapter provides an overview of crypto tokens and token offerings. Based on both utility tokens and security tokens, this chapter reviews the economics of tokens and token offerings. Specifically, it discusses the economic value of tokens for the financing, operations, and corporate governance of the issuing companies. It also discusses economic values for token investors. This chapter also discusses various token valuation models, as well as the underpricing and returns of the token markets.

1 Introduction

In this chapter, we provide an introduction to crypto tokens and token offerings. There are two main types of tokens: utility tokens and security tokens. Utility tokens give their holders access to product or service and that generally require the use of a blockchain-type infrastructure (Mougayar 2017; Fisch 2019; Yermack 2017). Security tokens are tradable tokens whose primary purpose is to give holders voting or financial rights and therefore mimic traditional financial assets such as debt and equity (Koffman 2018). Tokens represent assets and utilities of issuing companies and are issued to their investors in token offering events.

In the blockchain industry, initial coin offerings (ICOs) refer to the initial offering of utility tokens and security token offerings (STOs) are the initial offerings of security tokens (Blockgeeks 2018). In this chapter, for simplicity, we use ICOs to refer to both initial offerings of utility tokens and security tokens.

The chapter is organized as follows. It first provides an overview of tokens and token offerings, with discussions of various types of tokens, and a comparison between initial token offerings (ICOs) and initial public offerings (IPOs). It then

C. Liu (✉)
Trinity Western University, Langley, BC, Canada
e-mail: chen.liu@twu.ca

H. Wang
Coinchain Capital Inc., Vancouver, BC, Canada
e-mail: harry.wang@coincc.ca

S. Goutte et al. (eds.), *Cryptofinance and Mechanisms of Exchange*, Contributions to Management Science, https://doi.org/10.1007/978-3-030-30738-7_8

discusses the token economics, usually referred to as "tokenomics", which specifies the economics behind token offerings, the economic value of tokens for token issuers and investors, and corporate governance with tokens. The next section discusses valuation of crypto tokens, based on monetary theories and traditional valuation methods for equity and assets. The section afterwards discusses ICO underpricing and returns. The last section concludes the chapter.

2 Tokens and Token Offerings

2.1 What Are Crypto Tokens?

The first cryptocurrency, bitcoin, was created in 2009 from an anonymous white paper as a method of payments (Nakamoto 2008). Ethereum is an alternative currency to Bitcoin, developed in 2014, which enables automatically executable smart contracts (Buterin 2013). Tokens thereafter are created as smart contracts on top of blockchain, often based on the Ethereum network .

There are three main types of tokens: utility tokens, security tokens, and crypto-currency tokens.

Utility Tokens Tokens that confirm rights to access to product or service and that generally require the use of a blockchain-type infrastructure (Catalini and Gans 2017).

Security Tokens Tradable tokens whose primary purpose is to give holders voting rights and/or financial rights. Specifically, security tokens, also called tokenized securities or investment tokens (Koffman 2018), are financial securities compliant with security regulations and can provide financial rights to investors such as equity, dividends, profit sharing rights, and voting rights. Security tokens usually represent rights to underlying assets such as cash flow, real estate, and collectibles such as arts.

Compared to traditional debt and equity, advantages of security tokens include (1) fractionalization of larger assets, (2) increased liquidity as it is easier to get tokens listed on crypto exchanges compared to equity, (3) lower issuance fees compared to traditional equity and debt underwriting, (4) access to a global pool of capital and more market exposure as deals are so visible to everyone with internet connection (Koffman 2018; Malinova and Park 2018; Marks 2018).

Marks (2018) considers equity security tokens, security tokens that possess characteristics similar to equities, as one of the most promising crypto-asset classes. He argues that these tokens have some characteristics that make them better than traditional equities in certain ways. First, in theory, equity security tokens can be traded all year long on crypto exchanges or OTCs without any geographic or time limitations, contrary to traditional stocks. Second, specific terms such as vesting periods and investor restrictions of tokens can be easily designed and formulated in the smart contracts, which makes governance and management of these tokens less subject to manipulation (Yermack 2017).

Crypto-currency Tokens Tokens accepted as a means of payment for the purchase of goods and/or services, or to be used for the money or value transfer. Bitcoin, Bitcoin Cash, and Litecoin are examples of crypto-currency tokens. Crypto-currency tokens are independent of a particular platform and can be used as a form of currency outside their native environment, whereas utility tokens and security tokens in general exist on a particular platform that the token issuers create (Blockgeeks 2018).

For the purpose of this chapter in studying token offerings and tokenomics, we focus on the utility tokens and security tokens, as the issuance of these tokens are related to the real operation and/or finance of the token issuers (Gan et al. 2019; Momtaz 2019a).

One crucial step in a token offering is the Howey Test that lays down criteria according to which a token might be considered a security from a regulatory standpoint (Momtaz 2019a). The four main criteria of the Howey Test are (1) there is investment of money, (2) profits are expected, (3) money investment is a common enterprise, and (4) any profits come from the efforts of a promoted or third party. Most of the tokens, therefore, according to the Howey Test, would fall under the category of security tokens (Blockgeeks 2018).

Compared to issuance of utility tokens, it is more costly to issue security tokens as they are subject to greater security regulations and therefore a higher legal and disclosure costs. In the U.S., security tokens need to follow Regulation D, Regulation S, or Regulation A+ (Blockgeeks 2018). Nevertheless, security tokens act like a bridge between real assets and cash flows and the blockchain world.

Overall, the emergence of tokens and token offerings enables entrepreneurs to respond to two fundamental needs of the blockchain ecosystem. First, it creates incentive mechanisms to participate to this ecosystem and to innovate. Second, it provides the financial ability to fund the project, which allow entrepreneurs to fund their digital platform, software or other projects at an early stage of their development (Iansiti and Lakhani 2017).

2.2 A Comparison of Token Offerings and IPOs

Token offerings are essentially crowdfunding enabled by smart contracts for the purpose of funding blockchain-based companies or projects (Momtaz 2019a). As a financing strategy, ICOs are also frequently compared to IPOs of stocks (Liu 2019; Ofir and Sadeh 2019). In this subsection, we discuss the key differences between ICOs and IPOs.

The first difference lies in the type of securities issued. In an IPO, companies issue equity shares where investors realize returns through dividends and/or capital gains.

While security token offerings are similar to IPOs, utility tokens give their holders access to future product or service without directly sharing issuers' profits.

The second difference is the stage of the company. An IPO typically occurs at a later stage in a company's life cycle, where the company has viable product and/or service and earned revenue and is close to being profitable. An ICO in comparison is typically for a new, usually unproven concept that is seeking to raise capital (Liu 2019). Therefore, IPOs are usually for well-settled companies as exit strategies, whereas an ICO is more for young and risky companies to raise their initial financing (Liu 2019).

Third, IPOs are highly regulated, whereas ICOs are way less-regulated, or almost self-regulated, although some countries have tightened ICO regulation (Rhue 2018). Companies that issue their stock for the first time go through a complex IPO process, filing a lengthy IPO prospectus in order to get approved by security commission, while in early days ICOs companies often just disclose a whitepaper. Another major difference is the listing requirements—in order for an IPO to sell shares and thus provide liquidity to existing shareholders, it must be listed on an exchange. ICOs in comparison are not obligated to list on any cryptocurrency exchange, and in fact many ICO issuers fail to list on crypto exchanges (Momtaz 2019a).

The fourth main distinction is the investor type. In order to subscribe for an IPO, an investor must be deemed as sophisticated with basic requirements to be met. In fact, IPOs are often allocated only to institutional investors such as investment banks, mutual funds and endowments (Liu 2019). In an ICO of utility tokens, the investors are not known and there are in general no requirements on the investors' sophistication. For security tokens though, investors still need to be accredited investors, at least in the U.S. (Blockgeeks 2018).

3 The Emergence of Tokenization and Tokenomics

3.1 Tokenomics

ICO first started as entrepreneurs could not raise enough capital through traditional fundraising methods, therefore innovative ways of fundraising were necessary (Chen 2018). ICOs were then invented to create a more direct relationship between blockchain entrepreneurs and investors.

The importance of the ICO and tokens is to be understood through the economic functions of tokens, what is frequently coined as "tokenomics" (Malinova and Park 2018). Ennis et al. (2018) propose three definitions of tokenomics: "(1) a means of self-funding within the crypto economy, (2) the deployment of a token within the ecosystem of an ICO project and (3) the set of all economic activity generated through the creation of tokens". The first definition suggests the funding role of tokens, and the second and third definitions consider tokens as important incentives to use the technology provided by the token issuer, and in a broader sense, focus on the economic activity and value generated through the token creation.

The Network Effects and the Token Price

A strand of literature examines the network effects in blockchain-empowered token projects (Bakos and Halaburda 2018; Li and Mann 2018; Sockin and Xiong 2018). Sockin and Xiong (2018) model a platform token as the only currency accepted in a network. Li and Mann (2018) and Bakos and Halaburda (2018) highlight that most ICO projects are designed to create positive network effects that the token holders can monetize later. This is consistent with the second and third definitions of tokenomics of Ennis et al. (2018). Specifically, blockchain projects aim to create a network of users, often referred to as a "community", and tokens are used as an incentive mechanism to reward network contributors. A contributor can be an engineer who writes code for blockchain development, a financier who contributes fiat or cryptocurrency investments, or a community member who helps advertise and market the projects and token sales (Cong et al. 2018). Contributors are paid in tokens, and their inputs drive the quality of the blockchain platform. As the quality of the network improves, it is more attractive for users to buy tokens to gain access to the network, which further makes it even more attractive to contribute to the network (Klöhn et al. 2018). This relationship is intended to create positive network effects to make the network more attractive for all users.

Tokens, if well designed, will provide novel ways of incentivizing the network and monetizing network effects. Because tokens provide access to the network, their value correlates positively with the appeal of the network. The more attractive the network, the higher the demand for tokens, the higher the value of tokens. As long as tokens are kept scarce, a higher demand for tokens leads to a higher price of the tokens (Li and Mann 2018; Klöhn et al. 2018). Cong et al. (2018) formally model token valuation with the network effect. They argue that token transactions give token holders a flow utility that depends on tokenholder-specific needs, the size of the platform user base, and the platform quality.

Token holders can then sell their tokens on a secondary market, in exchange for cryptocurrencies such as Bitcoin or Ether or fiat money. This is the innovation of ICOs with a liquid secondary market for the tokens and thereby enabling token holders to monetize the network effect (Amsden and Schweizer 2018; Lee and Parlour 2019; Momtaz 2019a). This is particularly valuable for earlier contributors/investors, who are able to purchase the tokens at lower price (Catalini and Gans 2017). In addition, any increase in network value will immediately be reflected in the token price because tokens are scarce and are necessary to gain access to the network (Klöhn et al. 2018).

3.2 *The Economic Value of Tokens for Entrepreneurs*

3.2.1 Benefits of ICOs to Entrepreneurs

ICOs are an important innovation in entrepreneurial finance that have several advantages over traditional financing channels, particularly in mitigating moral hazards and asymmetric information (Momtaz 2019a; Howell et al. 2018).

First, significant information asymmetries exist in the traditional entrepreneurial finance that impedes entrepreneurs' access to capital. Specifically, traditionally, investors who wish to invest in high risk and high reward projects have little access to the projects' information, and entrepreneurs have few connections to such investors. A first improvement is through crowdfunding platform such as Kickstarter and Indiegogo, which presents startup projects on the Internet and thereby increasing the capital-raising opportunities for small business (Mollick 2014) and democratizing access to capital (Mollick and Robb 2016). Making information readily available on the Internet significantly reduces information asymmetries. A further improvement is made by ICOs: raising capital on the internet via blockchain technology connects entrepreneurs with a wide range of investors including future customers, thus reducing information frictions substantially (Adhami et al. 2018; Catalini and Gans 2017, 2019; Momtaz 2019a; Li and Mann 2018; Lipusch 2018). Importantly, after the tokens are listed on crypto exchanges, they provide liquidity for tokenholders, which is key advantage over private equity investment and crowdfunding (Lee and Parlour 2019).

A second significant benefit of ICOs is that since token sales are based on blockchain technology, issuers usually have to establish immutable and non-negotiable governance terms through smart contract (Howell et al. 2018). These terms are available to investors ex ante and are theoretically impossible to change ex post, signaling strong commitment of the founding team on governance (Yermack 2017).

Third, ICOs use decentralized networks, in which values generated in the network would accrue to its token holders. This is consistent with the network effect discussed above. Chen (2018) therefore argues that blockchain tokens give entrepreneurs new ways to engage key stakeholders and to develop, deploy, and diffuse decentralized applications. While an ICO can compensate initial investors and developers, it does not give them more control of the network than any other token holders (Garratt and van Oordt 2019; Howell et al. 2018). This helps alleviate the concern of moral hazard in traditional networks, where investors or customers worry the first-comers and developers extract rents from the network (Lee and Parlour 2019).

ICOs and token issuing are important features that facilitate the blockchain open source projects. A computer program is open source when its underlying source code is freely available, which means developers will not be rewarded from the project itself (Klöhn et al. 2018). Token sales solve this problem by creating an opportunity for developers to participate in the economic success of the project. If tokens are necessary to use the platform or services offered within the network, any increase in the value of the network is reflected in an increased demand and consequently a higher value of the tokens, which the developers can monetize via the sale of tokens on the secondary market. In addition, tokens give users an incentive to become an early contributor in the development of software, as they can directly profit from their contribution of value (Klöhn et al. 2018). Thus, the interests of the developers and other statekholders are aligned right from the start (Catalini and Gans 2017).

3.2.2 Drawbacks

Regardless of the benefits discussed above, ICOs have their own drawbacks. First, most ICOs only consists of one round of financing. The one-round-only design is necessary because the initial supply of tokens typically is fixed (Klöhn et al. 2018). However, this means ICO projects do not have the opportunity of further financing rounds as in angel or venture capital (VC) investments, which may limit the amount raised through ICOs over the long term.

Second, token sales can be tax inefficient (Cook and Heath 2017). The proceeds raised through token sales are treated as revenues or deferred revenues, which are subject to tax. In contrast, funds raised through equity financing are not treated as revenues and thus are not subject to tax.

An additional disadvantage is the regulatory uncertainty with ICOs. In 2017, some countries (e.g., China and South Korea) banned ICO (Choudhury 2017; O'Leary 2017). In the U.S., ICOs are not illegal, yet the Securities and Exchange Commission (SEC) has not offered clear guidelines regarding token sales.

3.2.3 Implications for Entrepreneurs

Researchers are starting to provide guidance to start-ups looking to issue tokens through ICOs (Conley 2017), including the technical factors and business elements that influence success. Although ICOs have the potential to disrupt the VC process (Lipusch 2018), there is not much guidance for entrepreneurs or investors on how to maximize this opportunity.

As a technical matter, as most tokens are created on smart contract, which is immutable once it is deployed, start-ups must choose the parameters of their token carefully. Prior to the token launch, entrepreneurs must identify a number of technical elements of the ICO such as the total supply of tokens, the token decimals, and the initial price.

In addition to the token details, companies must decide their business practices such as strategy, marketing, and issuing jurisdiction. To attract investors and provide information, token issuers often build their corporate websites, post their white paper, and share corporate information on social media. Value of the tokens are associated with the white paper quality and social media attention (Bourveau et al. 2018; Liu and Wang 2019).

Also, because ICOs are a global phenomenon (Zetzsche et al. 2018), token issuers must decide in which jurisdiction to issue their tokens. For instance, although the U.S. security law is unclear on the status of tokens at the time of writing (Rohr and Wright 2017), the U.S. government currently views the sale of tokens in the U.S. as a form of securities, requiring that companies vet their investors and/or verify the investor status as "accredited investors". In addition, token issuers who wish to accept investment from U.S. citizens must comply with U.S. know-your-customer (KYC) and anti-money laundering (AML) regulations and gather detailed about their customers. Companies that are KYC/AML-compliant may be more successful due to their access to U.S. investors and signaling of better quality (Lyandres et al. 2019).

3.3 The Economic Value of Tokens for Investors

Investors purchase tokens because they expect the underlying value of the tokens to increase, either through exchanging the tokens for goods and services or through its resale in a secondary market, either on a crypto exchange or over-the-counter (OTC) (Amsden and Schweizer 2018; Momtaz 2019a). Volatility of token prices in the secondary markets may attract investors looking for a high risk-return profile, with confident investors tempted by the prospect of identify the "next Bitcoin" (Masiak et al. 2018).

In addition to the financial reasons, Fisch et al. (2018) propose that investors invest in ICOs because they want to support the anonymity and decentralization of the blockchain system (the ideological reasons) and they value the technology of ICO firms (the technological reasons).

There are multiple ways to invest in ICOs (Colak and Hoogeveen 2017). In order to understand the value proposed by the cryptocurrency, the investor must read the white paper and research the company itself (Liu and Wang 2019). However, since the relationship between cryptocurrencies and traditional assets tend to be low, traditional analyses for security valuations may not be applicable (Bheemaiah and Collomb 2018). The next section will discuss more details on token price and valuation.

Rapid liquidity after ICO exchange listing is another benefit of ICOs (Momtaz 2019a; Howell et al. 2018). It permits a broader range of individuals, who may be excluded in traditional financing instruments, to invest in high-risk, high-return venture projects. In addition, crypto tokens, whether utility or security tokens, are a new asset class that allows investors to diversify their investment portfolios (Feng et al. 2018).

3.4 Corporate Governance with Tokens

Corporate governance is the way in which a corporation is directed, administered, and controlled (Baker and Anderson 2010). There are two ways tokens can impact corporate governance. First, token holders, as a new group of stakeholders, can affect the balance of power within companies. Second, tokens make possible a completely new governance structure, such as the decentralized autonomous organization (DAO).

3.4.1 Token Holders as a New Type of Corporate Stakeholders

Security Tokens and Corporate Governance

At the corporate governance level, the main issue for security token holders is to know whether they legally have and could exercise the ownership, cash flow or control rights granted to them. For example, Blemus and Guégan (2019) find that tokens could avoid granting voting rights or rights to liquidation surplus. In addition, it is still not clear whether the purchase of security tokens (during ICOs, on crypto-exchange platforms, by OTC transactions, or else) could have similar qualifications as the purchase of 'traditional' securities such as equity or debt instruments (Blemus and Guégan 2019; Marks 2018). There is also concern for market abuse, where token prices can be manipulated by not-yet-regulated crypto exchanges or investors with significant holdings, that could negatively affect the issuing companies (Keidar and Blemus 2018).

Utility Tokens and Corporate Governance

Blockchain entrepreneurs create utility tokens to raise funds without granting investors economic rights nor having any substantial fiduciary duty to the investors (Bheemaiah and Collomb 2018; Catalini and Gans 2017). While utility token holders have no control rights, the market value and trading volumes of these tokens would represent an important role in exerting pressure for the token holders to have an indirect impact on the company's decisions (Blemus and Guégan 2019; Yermack 2017). It is therefore important for the token holders to develop a direct dialogue with the corporation and to send requests to the company management. In the long term, companies will have to rethink the role of utility token holders and ways to develop interactions and communications with these new group of corporate stakeholders (Yermack 2017).

3.4.2 The Emergence of Distributed Governance

The DAO (Decentralized Autonomous Organizations)

The DAO (decentralized autonomous organizations) represents a new kind of organizations (Yermack 2017). Specifically, the DAO governance is based on a structure where the corporate decisions are decided by token holders' online voting processes (Buterin 2014; Chohan 2017; Jentzsch 2016). While the DAO fund later collapsed, it highlighted investors' willingness to support a new type of funding mechanism that is inherently built on anonymous trust and voting. It started a new decentralized/distributed form of corporate governance based on peer-to-peer cooperation and on consensus automated decision-making processes (Yermack 2017).

Distributed Organization Models

The use of blockchain technology, smart contracts, tokens and token offering has allowed many innovators to think about new models of corporate governance (Yermack 2017). Developing consensus mechanisms for corporate decisions could alter the fundamentals of corporate governance, such as the firm theory, the agency theory and the relationship between agents and principals (Jensen and Meckling 1976), beyond the traditional centralized and hierarchical governance structure of firms.

Some recent studies (such as De Filippi 2018; Feng et al. 2018; Fenwick and Vermeulen 2018; Johnson and Yi 2018; Wright and De Filippi 2015; Yermack 2017) have considered the distributed and consensus mechanisms of blockchain tokens as an instrument to solve corporate governance issues. ICOs can alleviate asymmetric information and incentive problems through self-imposed governance mechanism despite the limited regulation in the crypto market (Johnson and Yi 2018). The tokens and smart contracts could potentially provide a full and constant transparency and verifiability of the data available to key stakeholders for corporate management (Davidson et al. 2016). In this way of thinking, the replacement of trust in a disruptive technology management instead of trust in a human management team would be a strong incentive to minimize agency costs (De Filippi 2018; Yermack 2017).

4 Valuation of Crypto Tokens

The book of Burniske and Tatar (2017) is one of the first studies on crypto token valuation by underlying the similarities between stock and token valuation and applying the traditional valuation methods to crypto assets. They discuss traditional valuation methods such as the discounted cash flow (DCF) method, P/E ratio and the velocity of circulation. They therefore suggest that when examining a crypto asset, the fundamental analysis ought to include: (1) whitepaper, (2) technical aspects (e.g., hash rate, number of miners), (3) community and developers, (4) relation to other crypto assets, and (5) issuance model.

Since the book, there has been a growing interest in examining the valuation methods for tokens, including studies based on the traditional monetary theory (Buterin 2017; Weber 2018) and new terms such as Crypto J-Curve (Burniske 2017). The rest of this section discusses each valuation method. Studies mentioned here are mostly from practitioners' side and the academic studies (e.g. Cong et al. 2018; Pazos 2018, 2019) are catching up lately.

4.1 *Token Velocity Methodology*

The token velocity methodology applies the Quantity Theory of Money (QTM) to a token-based economy (Buterin 2017; Weber 2018). It has therefore gained a lot of ground in the discussion of utility tokens valuation.

Specifically, the QTM states that the general price level of goods and services is directly proportional to the amount of money in circulation, or money supply

(Friedman 1956). The QTM is based on the definitional relationship: MV = PQ, whereas M indicates the money supply in the economy, V is the velocity of circulation, P is the price level, and Q is the output produced by the economy. Applying it to tokens, we have the following equation:

$$MV = PQ$$

$$Token\ Price = \frac{1}{P} = \frac{Q}{MV}$$

whereas

- M is the total number of tokens
- V is token velocity, that is, the number of times that an average token changes hands
- P is the price of goods and services in terms of the token, and therefore it is the inverse of the token price
- Q is the economic value of token transaction per day

The method hence states that velocity is one of the more important drivers and indicators of valuation (Evans 2018; Lannquist 2018; Weber 2018). The implication is that tokens with low velocity, i.e., those that held (owning to speculation, asset backed, and etc.), will see prices rise (Bheemaiah and Collomb 2018).

This valuation methods can be applied to both the general purpose cryptocurrencies such as the Bitcoin and the utility tokens used in a smart contract platforms (Bheemaiah and Collomb 2018). The reasoning behind this approach is that as the token of a smart contract platform becomes widespread and sufficiently useful, it will emerge as an independent store of value (Samani 2018).

4.2 *Crypto J-Curve Methodology*

Burniske (2017) proposes the Crypto J-Curve. While J-Curve in economics is used to describe the effects of currency devaluation on the national deficit, and in private equity refers to a portfolio's cash flow, Burniske (2017) uses the J-Curve to capture the market values of crypto assets over time. Specifically, a token's price is composed of two forms of value: (1) "current utility value" (CUV), which represents value driven by utility and usage today, and (2) "discounted expected utility value" (DEUV), which represents value driven by investment speculation for the future (Burniske 2017).

According to Burniske (2017), CUV and DEUV take turns driving token prices as a blockchain project develops and its market perceptions change accordingly. Specifically, when a project and its token are first launched, CUV is low and DEUV dominates as holders are excited about the technology and expect future price appreciation. When enthusiasm wanes and DEUV drops with inevitable technical roadblocks, token price drops and is driven more by CUV from the project's early adopters. As the team overcomes challenges, CUV grows as the token becomes more widely adopted, driving up the token price. DEUV then catches up as speculation and excitement start to grow again. Ultimately in the steady state of the blockchain project, CUV should drive token price.

Linking back to stock valuation, the notion of DEUV can be considered as a modified version of the DCF valuation method. Instead of measuring expected future cash flow, this model is a first step in estimating CUV and DEUV and their respective dynamic influences on token price (Bheemaiah and Collomb 2018).

Some adopters of the Crypto J-curve have begun to use it as a proxy for measuring the different life stages of a cryptoasset. For example, a New York based VC investment fund, Placeholder uses the curve to determine which stage a token sale is at: a whitepaper stage is where the team works to define and implement a "minimum viable protocol" and to validates the network's functionality, a release stage is when a token is first made available to the public, and a public stage when the token begins trading on exchanges (Monegro and Burniske 2017).

4.3 Network Value-to-Transaction Ratio (NVT)

In traditional stock markets, price-earnings ratio (P/E ratio) has been a long standing tool for equity valuation. A high P/E ratio indicates either over valuation or a company in high growth. Applying the P/E ratio to the crypto world, Woo (2017) suggests using money flowing through a token's network as a proxy to "earnings", leading to the NVT (network value to transaction ratio) method of token valuation:

$$\text{NVT (network value to transaction ratio)} = \text{network value/daily transaction volume.}$$

This valuation ratio compares the network's value (the market cap) to the network's daily on-chain transaction volume. Similar to the P/E ratio, the NVT may indicate whether a network token is under or overvalued by showing the market cap relative to the network's transaction volume, which represents the utility that users derive from the network. When the ratio becomes very high, it indicates potential token over-valuation.

The NVT methodology is consistent with the network theory of the tokens discussed above, as it emphasizes the overall utility of the network. Moving forward, using NVT will require some formal definition on what constitutes a valid transaction in certain networks.

4.4 Security Token Valuation

The methods discussed above are primarily related to the evaluation of utility tokens. When it comes to security tokens, the valuation models are more traditional as they are financial securities, providing an array of financial rights to investors such as equity, dividends, profit share rights, voting rights, etc. (Koffman 2018). While moving securities onto a Blockchain can have advantages in comparison to a legacy system in terms of settlement times, lower fees, automated service functions and custodianship, this does not change anything about the nature of the security itself (Bheemaiah and Collomb 2018). Hence, evaluation models of traditional securities, such as the DCF valuation, relative methods (e.g., P/E), or option pricing model, can be applied to valuation of security tokens.

4.5 Traditional Valuation Methods in Crypto Valuation

In this subsection, we further discuss whether and how traditional valuation methods can be used in token valuation in general.

Crypto CAPM

It would be interesting to explore how a multi-factor CAPM model could be applied to crypto asset valuation. Lannquist (2018) suggests using the following factors in a crypto multi-factor CAPM model:

- Momentum factor
- Liquidity factor (potentially measured by trading volume, bid/ask spreads, or small-cap minus large-cap returns as in CAPM)
- Token exchange and storage frictions (prevalence on centralized exchanges and decentralized exchange protocols, convenience to purchase, wallet quality, etc.)
- Community size/strength factor
- Value: low NVT vs. high NVT factor
- "FOMO" factor (beware of multicollinearity w/momentum and other factors)
- Global political or economic uncertainty

Since historic return periods are short, the model will be more effective in the future when the crypto asset markets mature and we have more data to study the relationship of token price and its various drivers.

Discounted Cash Flow Analysis (DCF)

Generally speaking, DCF is not suitable for utility tokens because they do not generate cash flows or represent equity claims on cash flows. However, a DCF valuation would be a great tool to value security tokens that provide equity features such as expected dividends or distributions.

Comparables Valuation Approach
In traditional equity valuation, the financial ratios and multiples of comparable companies can be used to imply share prices for a target company. Multiples such as P/E, EV/EBITDA, EV/Sales are applicable for security tokens and methods with token-relevant metrics such as NVT can be applied to utility tokens.

In summary, crypto markets are very new with limited data history pertaining to crypto asset behavior, returns, and correlations (Lannquist 2018). Many of today's models are simplistic or limited. In the future, when the markets mature and asset relationships and behaviors are more discoverable, valuation models should be more predictive and informative. As crypto assets are an emerging alternative asset class, much work is yet to be done studying valuation frameworks that can help investors estimate token prices. This calls for serious future research in the crypto area.

5 ICO Underpricing and Token Returns

5.1 Underpricing and First-Day Returns

Underpricing is the phenomenon whereby the price of an asset is set too low on issuance. As a result, the price adjusts to its market value on the listing day and underpricing is indicated by a large first day return (Loughran and Ritter 2002). Empirical studies find a significant evidence for underpricing in ICOs. For instance, Adhami et al. (2018) find that the mean (median) value of first-day return is 929.9% (24.7%). Benedetti and Kostovetsky (2018) find that the average first-day returns to be 179%. Bourveau et al. (2018) document the mean (median) first-day return to be 39% (40%). Overall, these results are evidence of significant underpricing in ICOs, although the degree of underpricing differs depending on the ICO sample and sample period.

Some studies offer a theoretical explanation for ICO underpricing, mostly in line with IPO underpricing. Momtaz (2019b), for example, argues that ICOs have an incentive to underprice their token to attract a large user base, which is an important signal for investors in particular with large degree of information asymmetry in ICOs. Benedetti and Kostovetsky (2018) argue that the information asymmetry associated with the market, coupled with the projects' early stages of development during the offering, are the main reasons for this underpricing. Similarly, Howell et al. (2018) suggest that in the absence of measures of commercial success, liquidity is a major signal of ICO quality from early investors' perspective. Cong et al. (2018)'s network model argues that when a platform has a token investors (users) join the platform, they not only enjoy its token utility, but also benefit from the rising token price as a result of the growing network size.

Momtaz (2019a), Benedetti and Kostovetsky (2018), Lyandres et al. (2019) and Felix and von Eije (2019) analyze the determinants of ICO underpricing. Benedetti and Kostovetsky (2018) and Felix and von Eije (2019) find that presales have a significant negative influence on underpricing. This result is consistent with Howell et al. (2018) and Lee et al. (2018)'s argument that early investment rounds provide an indication of the demand for the token, thus helping determine an appropriate price for the launch of the ICO. Felix and von Eije (2019) and Lyandres et al. (2019) find that the issue size of an ICO is negatively associated with underpricing, indicating that larger ICOs are associated with a lower degree of information asymmetry. They suggest that successful presales generate an information cascade during the launch of the ICO, encouraging subsequent investors to invest regardless of their own information

Conversely, Momtaz (2019a) finds that issue size is positively associated with ICO underpricing. Momtaz (2019a) also finds that country restrictions are positively associated with ICO underpricing, suggesting that higher incentives are required for the remaining potential investors. Interestingly, in contrast with IPOs, Chanson et al. (2018) and Benedetti and Kostovetsky (2018) find no significant association between firm's age and underpricing. Even though this may look surprising as older companies have had more time to reduce information asymmetry, in ICO markets, issuing companies are in general young startups, and therefore the firm age effect may not be at play here.

5.2 *Long-Term Returns and Performance*

Regarding the long-term return, most empirical studies find that the average long-term returns are usually positive, with a median number being negative. For instance, using a sample of ICOs between 2013 and January 2018, Howell et al. (2018) analyze the return between the first day of trading and 5 months later relative to the Bitcoin benchmark. They find that the average token price increases by 149% in this period, but the median decreases by 50%. Lyandres et al. (2019) find that the mean post-ICO cumulative return ranges between 6% for the 30-day and 365-day horizons to 46% for the 180-day horizon, but the median return is negative for all horizons, ranging from −29% to −78% with 67% (77%) of 30-day (365-day) cumulative returns being negative. These results are in line with Bourveau et al. (2018) who find a positive (39%) mean return for the 30-day horizon but the median value is negative (−30%). They also find a strong and positive correlation between first-day return and extreme negative return in the following 3–12 month period.

Momtaz (2019b) find that for a holding period between 1 and 24 months, the median ICO depreciates by 30% with substantial positive skewness. His results show that although there is significant ICO underpricing, 40% of ICOs are overpriced. He argues a size effect that large ICOs are more often overpriced and underperform in the long run. Interestingly, EY (2018) analyze the returns of 2017 ICOs from January to September 2018, and find that 86% of the ICOs were below listing price, and 30% lost substantially all their value. Hu et al. (2018) study the secondary market return of 222 tokens and find them to be strongly correlated with Bitcoin returns, suggesting that the return of Bitcoin itself is a primary risk factor in the crypto market.

5.3 Behavioral Biases in ICOs

Several empirical studies analyze behavioral biases of investor' sentiments, herding behavior, and speculative bubbles in the context of ICOs. First, consistent with the IPO literature, empirical studies document significant relationship between investors' sentiment and ICO market performance. Felix and von Eije (2019) find that market sentiment is positively associated with underpricing. Lee et al. (2018) find that first-day returns, as well as 1-week, 1-month and 3-month returns, are positively associated with the parallel market returns, suggesting that a hot crypto market increases investors' sentiment. Consistently, Momtaz et al. (2019) find that market sentiments and market liquidity are strongly associated with listing, suggesting that ventures have an incentive to conduct an ICO during hot crypto markets.

Other studies examine the influence of Ether and Bitcoin prices and volatility on ICOs. Masiak et al. (2018) find that shocks to Ether and Bitcoin affect ICOs, with shocks to Ether having a stronger effect. They also find that shocks to ICOs, as well as to Bitcoin and Ether, are persistent—a bullish market in ICOs remains bullish for 4 weeks. Momtaz (2019a) finds that Bitcoin price is positively associated with the amount raised and with first-day returns. Bourveau et al. (2018) find that past returns in Bitcoin, are positively associated with extreme negative returns in the following 3, 6 and 12 months, suggesting that issuers may strategically time their fundraising to hot markets and engage in "pump and dump" strategies that could harm investors.

Empirical studies also find evidence of herding in the crypto market. Calderón (2018) finds that herding behavior exists in the ICO market when the market exhibits positive returns, but reverses when it exhibits negative returns. Bouri et al. (2018) find that uncertainty, measured by the Economic Policy Uncertainty Index, increases the probability of herding. Their findings suggest that in the presence of market uncertainty, traders become more confident about the (upward) direction of cryptocurrencies and thus tend to mimic the trading actions. Overall, there is clear evidence for herding behavior in cryptocurrencies. These results are important as the herding phenomenon suggests that the efficient market hypothesis that assumes that investors trade rationally does not apply.

Sherman (2018) discusses the speculative bubbles in the ICO market. Speculative bubbles are defined as "unsustainable increases in asset prices caused by investors trading on a pattern of price increases rather than information on fundamental values" (Gerding 2007). In a bubble, informed investors "bid up prices in anticipation of 'noise traders' entering the market. The noise traders then enter the market due to the psychological biases they encounter in making their investment decisions" (Gerding 2007). Sherman (2018) and Bianchetti et al. (2018) find evidence of bubbles in cryptocurrencies in 2017 as "investors pour large amounts of money into the ICOs and the prices of coins issued in ICOs are only rising because other investors also funnel money into them".

6 Conclusion

This chapter provides an overview of crypto tokens and token offerings. Based on both utility tokens and security tokens, this chapter reviews the economics of tokens and token offerings. Specifically, it discusses the economic value of tokens for the financing, operations, and corporate governance of the issuing companies. It also discusses economic values for token investors. This chapter then discusses various token valuation models, as well as the underpricing and returns of the token markets. Discussions of this chapter provide insights for crypto-entrepreneurs, academics and regulators worldwide to better understand tokens and their economic values to various functions in companies and to investors.

References

Adhami S, Giudici G, Martinazzi S (2018) Why do businesses go crypto? An empirical analysis of initial coin offerings. J Econ Bus 100:64–75

Amsden R, Schweizer D (2018) Are blockchain crowdsales the new "gold rush"? Success determinants of initial coin offerings. https://papers.ssrn.com/sol3/papers.cfm?abstract_id=3163849

Baker HK, Anderson R (2010) An overview of corporate governance. Corporate governance: a synthesis of theory, research, and practice, pp 3–17

Bakos Y, Halaburda H (2018) The role of cryptographic tokens and ICOs in fostering platform adoption. https://papers.ssrn.com/sol3/papers.cfm?abstract_id=3426940

Benedetti H, Kostovetsky L (2018) DigitalE tulips? Returns to investors in initial coin offerings. Returns to Investors in Initial Coin Offerings (May 20, 2018)

Bheemaiah K, Collomb A (2018) Cryptoasset valuation. https://www.louisbachelier.org/wp-content/uploads/2018/10/cryptovaluationreport-v20181016-vf.pdf

Bianchetti M, Ricci C, Scaringi M (2018) Are cryptocurrencies real financial bubbles? Evidence from quantitative analyses. Evidence from quantitative analyses (February 24, 2018). A version of this paper was published in Risk, 26

Blemus S, Guégan D (2019) Initial Crypto-asset Offerings (ICOs), tokenization and corporate governance. Tokenization and Corporate Governance (January 11, 2019)

Blockgeeks (2018) What are security tokens? https://blockgeeks.com/guides/security-tokens/

Bouri E, Gupta R, Roubaud D (2018) Herding behaviour in cryptocurrencies. Financ Res Lett 29:216–221

Bourveau T, De George ET, Ellahie A, Macciocchi D (2018) Initial coin offerings: early evidence on the role of disclosure in the unregulated crypto market. https://www.marshall.usc.edu/sites/default/files/2019-03/thomas_bourveau_icos.pdf

Burniske C (2017) The crypto j-curve. https://medium.com/@cburniske/the-crypto-j-curve-be5fdddafa26

Burniske C, Tatar J (2017) Cryptoassets: the innovative investor's guide to bitcoin and beyond. McGraw Hill Professional

Buterin V (2013) Ethereum white paper. GitHub repository, pp 22–23

Buterin V (2014) On public and private blockchains. Etherium blog. https://blog.ethereum.org/2015/08/07/on-public-and-private-blockchains/

Buterin V (2017) Ethereum research update. https://blog.ethereum.org/2016/12/04/ethereum-research-update

Calderón O (2018) Herding behavior in cryptocurrency markets. arXiv preprint arXiv:1806.11348

Catalini C, Gans JS (2017) Some simple economics of the blockchain, Working Paper, University of Toronto

Catalini C, Gans JS (2019) Initial coin offerings and the value of crypto tokens. https://www.nber.org/papers/w24418

Chanson M, Gjoen J, Risius M, Wortmann F (2018) Initial coin offerings (ICOs): the role of social media for organizational legitimacy and underpricing. https://www.alexandria.unisg.ch/255399/

Chen Y (2018) Blockchain tokens and the potential democratization of entrepreneurship and innovation. Bus Horiz 61(4):567–575

Chohan UW (2017) The decentralized autonomous organization and governance issues. https://papers.ssrn.com/sol3/papers.cfm?abstract_id=3082055

Choudhury SR (2017) China bans companies from raising money through ICOs, asks local regulators to inspect 60 major platforms. https://www.cnbc.com/2017/09/04/chinese-icos-china-bans-fundraising-through-initial-coin-offerings-report-says.html

Colak S, Hoogeveen M (2017) TACXE: a blockchain based token listing and exchange platform. https://iceclog.com/wp-content/uploads/2017/11/TACXE-Position-Paper-v07nov2017.pdf

Cong LW, Li Y, Wang N (2018) Tokenomics: dynamic adoption and valuation. https://papers.ssrn.com/sol3/papers.cfm?abstract_id=3222802

Conley JP (2017) Blockchain and the economics of crypto-tokens and initial coin offerings. http://www.accessecon.com/Pubs/VUECON/VUECON-17-00008.pdf

Cook J, Heath M (2017) ICOs: compelling advantages, real risk, September 9, 2017. https://www.coindesk.com/token-sales-compelling-advantages-real-risk/

Davidson S, De Filippi P, Potts J (2016) Disrupting governance: the new institutional economics of distributed ledger technology. https://papers.ssrn.com/sol3/papers.cfm?abstract_id=2811995

De Filippi PDF (2018) Blockchain and the law: the rule of code. Harvard University Press, Cambridge

Ennis P, Waugh J, Weaver W (2018) Three definitions of tokenomics. https://www.coindesk.com/three-definitions-tokenomics

Evans A (2018) On value, velocity, and monetary theory: a new approach to cryptoasset valuations. https://medium.com/blockchannel/on-value-velocity-and-monetary-theory-a-new-approach-to-cryptoasset-valuations-32c9b22e3b6f

EY (2018) EY study: Initial Coin Offerings (ICO) The class of 2017—one year later. https://www.ey.com/Publication/vwLUAssets/ey-study-ico-research/$FILE/ey-study-ico-research.pdf

Felix TH, von Eije H (2019) Underpricing in the cryptocurrency world: evidence from initial coin offerings. Manag Financ 45:563–578

Feng C, Li N, Lu B, Wong MH, Zhang M (2018) Initial coin offerings, blockchain technology, and voluntary disclosures. https://papers.ssrn.com/sol3/papers.cfm?abstract_id=3256289

Fenwick M, Vermeulen EP (2018) Technology and corporate governance: blockchain, crypto, and artificial intelligence. Lex Research Topics in Corporate Law & Economics Working Paper (2018-7)

Fisch C (2019) Initial coin offerings (ICOs) to finance new ventures: an exploratory study. J Bus Ventur 34(1):1–22

Fisch C, Masiak C, Vismara S, Block JH (2018) Motives to invest in initial coin offerings (ICOs). https://papers.ssrn.com/sol3/papers.cfm?abstract_id=3287046

Friedman M (1956) The quantity theory of money: a restatement. Studies in the quantity theory of money. University of Chicago Press, Chicago, p 5

Gan JR, Tsoukalas G, Netessine S (2019) Inventory, speculators and initial coin offerings. The Wharton School Research Paper

Garratt R, van Oordt MR (2019) Entrepreneurial incentives and the role of initial coin offerings. https://papers.ssrn.com/sol3/papers.cfm?abstract_id=3334166

Gerding EF (2007) Laws against bubbles: an experimental-asset-market approach to analyzing financial regulation. Wis L REv 977

Howell ST, Niessner M, Yermack D (2018) Initial coin offerings: financing growth with cryptocurrency token sales. https://www.nber.org/papers/w24774

Hu A, Parlour CA, Rajan U (2018) Cryptocurrencies: stylized facts on a new investible instrument. https://papers.ssrn.com/sol3/papers.cfm?abstract_id=3182113

Iansiti M, Lakhani KR (2017) The truth about blockchain. Harv Bus Rev 95(1):118–127. https://hbr.org/2017/01/the-truth-about-blockchain

Jensen MC, Meckling WH (1976) Theory of the firm: managerial behavior, agency costs and ownership structure. J Financ Econ 3(4):305–360

Jentzsch C (2016) Decentralized autonomous organization to automate governance. White paper, November

Johnson WC, Yi S (2018) Governance in the absence of regulation: a study of initial coin offerings. https://papers.ssrn.com/sol3/papers.cfm?abstract_id=3337096

Keidar R, Blemus S (2018) Cryptocurrencies and market abuse risks: it's time for self-regulation. Lexology. https://papers.ssrn.com/sol3/papers.cfm?abstract_id=3123881

Klöhn L, Parhofer N, Resas D (2018) Initial coin offerings (ICOs): economics and regulation. https://papers.ssrn.com/sol3/papers.cfm?abstract_id=3290882

Koffman K (2018) Your official guide to the security token ecosystem. https://medium.com/@tatianakoffman/your-official-guide-to-the-security-token-ecosystem-61a805673db7

Lannquist A (2018) Today's crypto asset valuation frameworks. https://blockchainatberkeley.blog/todays-cryptoasset-valuation-frameworks-573a38eda27e

Lee J, Parlour CA (2019) Consumers as financiers: crowdfunding, initial coin offerings and consumer surplus. https://www.chapman.edu/research/institutes-and-centers/economic-science-institute/_files/ifree-papers-and-photos/parlour-lee-consumers-as-financiers-2019.pdf

Lee J, Li T, Shin D (2018) The wisdom of crowds and information cascades in FinTech: evidence from initial coin offerings. https://papers.ssrn.com/sol3/papers.cfm?abstract_id=3195877

Li J, Mann W (2018) Initial coin offering and platform building. https://papers.ssrn.com/sol3/papers.cfm?abstract_id=3088726

Lipusch N (2018) Initial coin offerings – a paradigm shift in funding disruptive innovation. https://papers.ssrn.com/sol3/papers.cfm?abstract_id=3148181

Liu C (2019) FinTech and its disruption to financial institutions. In: Organizational transformation and managing innovation in the fourth industrial revolution. IGI Global, pp 104–124

Liu C, Wang H (2019) Initial coin offerings: what do we know and what are the success factors? In: Goutte S, Guesmi K, Saadi S (eds) Handbook: Cryptofinance and mechanism of exchange. Forthcoming

Loughran T, Ritter JR (2002) Why don't issuers get upset about leaving money on the table in IPOs? Rev Financ Stud 15(2):413–444

Lyandres E, Palazzo B, Rabetti D (2019) Are tokens securities? An anatomy of initial coin offerings. http://abfer.org/media/abfer-events-2019/annual-conference/corporate-finance/30-P1-Are-Tokens-Securities.pdf

Malinova K, Park A (2018) Tokenomics: when tokens beat equity. https://papers.ssrn.com/sol3/papers.cfm?abstract_id=3286825

Marks H (2018) The future of US securities will be tokenized. https://hackernoon.com/the-future-of-us-securities-will-be-tokenized-c469d41d81a1

Masiak C, Block JH, Masiak T, Neuenkirch M, Pielen K (2018) The market cycles of ICOs, bitcoin, and Ether. https://papers.ssrn.com/sol3/papers.cfm?abstract_id=3198694

Mollick E (2014) The dynamics of crowdfunding: an exploratory study. J Bus Ventur 29(1):1–16

Mollick E, Robb A (2016) Democratizing innovation and capital access: the role of crowdfunding. Calif Manage Rev 58(2):72–87

Momtaz P (2019a) Initial coin offerings. https://papers.ssrn.com/sol3/papers.cfm?abstract_id=3166709

Momtaz P (2019b) The pricing and performance of cryptocurrency. Eur J Financ:1–14

Momtaz P, Drobetz W, Schroeder H (2019) Investor sentiment and initial coin offerings. J Altern Invest 21(4):26–40

Monegro J, Burniske C (2017) Placeholder: thesis summary. https://ipfs.io/ipfs/QmZL4eT1gxnE168Pmw3KyejW6fUfMNzMgeKMgcWJUfYGRj/Placeholder%20Thesis%20Summary.pdf?ref=tokendaily

Mougayar W (2017) Tokenomics—a business guide to token usage, utility, and value. https://medium.com/@wmougayar/tokenomics-a-business-guide-to-token-usage-utility-and-value-b19242053416

Nakamoto S (2008). Bitcoin: a peer-to-peer electronic cash system. https://bitcoin.org/bitcoin.pdf

O'Leary RR (2017) South Korean regulator issues ICO ban. CoinDesk. https://www.coindesk.com/south-korean-regulator-issues-ico-ban/

Ofir M, Sadeh I (2019) ICO vs IPO: empirical findings, market frictions and the appropriate regulatory framework. Market Frictions and the Appropriate Regulatory Framework (February 19, 2019)

Pazos J (2018) Valuation of utility tokens based on the quantity theory of money. J Br Blockchain Assoc 12382018(1):4318. https://doi.org/10.31585/jbba-1-2-(2)2018

Pazos J (2019) Valuation method of equity-based security token offerings (sto) for start-up companies. JBBA

Rhue L (2018) Trust is all you need: an empirical exploration of initial coin offerings (ICOs) and ICO reputation scores. https://papers.ssrn.com/sol3/papers.cfm?abstract_id=3179723

Rohr J, Wright A (2017) Blockchain-based token sales, initial coin offerings, and the democratization of public capital markets. Cardozo Legal Studies Research Paper, 527

Samani K (2018) New models for utility tokens. https://multicoin.capital/2018/02/13/new-modelsutility-tokens

Sherman NJ (2018) A behavioral economics approach to regulating initial coin offerings. Geo LJ Online 107:17

Sockin M, Xiong W (2018) A model of cryptocurrencies. https://wxiong.mycpanel.princeton.edu/papers/Crypto.pdf

Weber W (2018) The quantity theory of money for tokens. https://blog.coinfund.io/the-quantity-theory-of-money-for-tokens-dbfbc5472423

Woo W (2017) Is bitcoin in a bubble? Check the NVT ratio. https://www.forbes.com/sites/wwoo/2017/09/29/is-bitcoin-in-a-bubble-check-the-nvt-ratio/#18bdb3bf6a23

Wright A, De Filippi P (2015) Decentralized blockchain technology and the rise of lex cryptographia. https://papers.ssrn.com/sol3/papers.cfm?abstract_id=2580664

Yermack D (2017) Corporate governance and blockchains. Rev Financ 21(1):7–31

Zetzsche DA, Buckley RP, Arner DW, Föhr L (2018) The ICO gold rush: it's a scam, it's a bubble, it's a super challenge for regulators. Harv Int Law J 63(2):17–83

Initial Coin Offerings: What Do We Know and What Are the Success Factors?

Chen Liu and Haoquan Wang

Abstract This chapter reviews empirical studies on the characteristics of initial coin offerings (ICO) and determinants of ICO success. This chapter contributes to the literature by providing a discussion on all key elements in a full-cycle ICO and conducts comprehensive literature review of the common practice and key success factors for ICOs. Findings of this chapter provide important managerial and policy implications. Regulators should pay attention to the specific market frictions discussed in this chapter in order to provide a regulation framework that protects investors and promotes the market efficiency. The optimal regulatory framework should address information asymmetry by using disclosure provisions and impose legal obligations on analysts reviewing ICOs and the marketing materials.

1 Introduction

An Initial Coin Offering (ICO) is a fundraising mechanism for blockchain-related companies by issuing crypto tokens ("tokens" thereafter) (Boreiko and Sahdev 2018; Chod and Lyandres 2018). In general, there are two types of tokens: utility tokens that provide access to service or product the issuers will provide (Momtaz 2019a) and security tokens that grant their holders financial rights, such as dividends and voting rights (Collomb et al. 2018; Rohr and Wright 2017), debt-like rights (Barsan 2017), and real assets such as arts and real estate (Krypital Group 2018). While ICOs generally refer to the issuance of utility tokens and the blockchain industry usually uses the term security token offerings (STOs) for the issuance of security tokens, we follow Liu and Wang (2019) to use the term "ICOs" to refer to issuance of both utility and security tokens.

C. Liu
Trinity Western University, Langley, BC, Canada
e-mail: chen.liu@twu.ca

H. Wang (✉)
Coinchain Capital Inc., Vancouver, BC, Canada
e-mail: harry.wang@coincc.ca

S. Goutte et al. (eds.), *Cryptofinance and Mechanisms of Exchange*, Contributions to Management Science, https://doi.org/10.1007/978-3-030-30738-7_9

After the ICO, the tokens can be used to claim the product or service (in case of utility tokens) or converted into other cryptocurrencies or fiat money on a cryptocurrency exchange or over the counter (OTC). ICOs resemble IPOs as in both cases a company issues digital tokens or shares to raise capital, which is then publically traded (Liu 2019; Ofir and Sadeh 2019). ICOs also bear similarities to crowdfunding, as both methods allow startups and entrepreneurs to finance their project through the Internet, outside the traditional financing channels (Ante et al. 2018; Lee and Parlour 2019).

In this chapter, we review the ICO market and determinants of ICO success. We define ICO success as reaching fund-raising target, successfully listing tokens on crypto exchanges, or generating positive returns after the ICO. The purpose of this chapter is to provide an exploratory study that deepens our understanding on ICOs and to draw important managerial and policy implications for both practitioners and regulators.

The rest of the chapter is organized as follows. It first provides an overview of the ICO market. It then discusses two key stages of a full-cycle ICO process—the preparation stage and the issuance stage. In the preparation stage, issuers design the business model and token model, prepare the white paper (similar to an IPO prospectus) (Ante et al. 2018), deal with the technical and legal aspects, and conduct marketing and manage media relation. The ICO issuance starts from a presale, followed by a full ICO launch, and then a post-ICO token management, including exchange listing and market making. In each key stage, we discuss current practice and how they relate to ICO success.

2 The Market for ICOs and Cryptocurrencies

2.1 ICO Market Overview

ICOs have revolutionized the way startups fund their growth. The ICO market has experienced a rapid growth since its early days in 2014, with 1070 ICOs raised over $21 billion in 2018 (CoinSchedule 2018). Studies have offered various explanations for the rapid growth of the crypto market and ICOs in particular. Some argue that cryptocurrencies are perceived by investors as a “hedge against volatile local currencies and geopolitical risk”, and their growth is related to a continuing distrust in the traditional banking sector since the 2008 financial crises (Clements 2018). Other reasons include the increased media attention (Clements 2018) and extraordinarily high returns for early investors, with ROIs exceeding 50,000% (Hacker Moon 2017).

The crypto market has slowed down significantly in the latter half of 2018: with the aggregate cryptocurrency market falling by over 85% from its peak within a few months and funding decreasing by over 90% between July 2018 and February 2019 (Dittmar and Wu 2019). Fundraising success rate is also decreasing. Empirical studies find that ICO fundraising success dropped sharply since the second half of

2017 (Lee et al. 2018). Adhami et al. (2018) find an 81% success rate for ICOs that occurred from 2014 to August 2017. EY (2017) find that 90% of the ICOs reached their fundraising goals in June of 2017 but the number dropped to less than 25% in November of 2017. Benedetti and Kostovetsky (2018) find that for a sample of 2390 ICOs from January 2017 to March 2018 only 48% had capital raised and that only 26% have listed their tokens on crypto exchanges. Lee et al. (2018) attribute this decline in the ICO market to increased regulation and the drop of price of major cryptocurrencies, mostly Bitcoin and Ether.

Some empirical studies examine the geography of ICOs and find some different results depending on their sample and time period of study. For instance, Amsden and Schweizer (2018) find that the top country in the number of ICOs is the US, followed by Russia and that the dominant country in amounts raised is the US for $2.4 billion, followed by Switzerland of $1.1 billion. Huang et al. (2018) find that the top countries in the number of ICOs are the US, China, Russia, and Switzerland, and in amounts raised, the US, Switzerland, Singapore, and Russia.

2.2 *Four Stages of ICOs*

Based on the authors' experience, we summarize that a full-cycle ICO is mostly composed of four main stages. An ICO starts with an initial planning stage that includes strategic planning for the project, preparation of ICO documents such as white paper and presentation deck, token design and development, technology development, marketing, and legal preparation. Most ICOs will have a pre-ICO stage, in which the token issuers sell tokens to early investors at discount. Then comes the full ICO launch, usually lasts for 1–3 months or until the hard cap is reached. After the ICO, issuers will get the token listed on crypto exchanges and conduct market making to provide liquidity. Also, during this post-ICO stage, token issuers use the mount raised to develop blockchain projects.

In recent years, there has been a growing body of empirical and theoretical studies that analyze determinants of ICO success. Results are often inconsistent, mostly due to sample differences and the ever-changing crypto market conditions. In the next two sections, we review existing studies on ICOs by examining stages of ICOs. Understanding the process is essential both for analyzing determinants of ICO success and for designing an optimal regulation regime.

3 Planning Stage of ICOs

3.1 *White Paper*

The main document in an ICO is a white paper, similar to an IPO perspective (Barsan 2017). According to Bourveau et al. (2018) and the authors' own experience, while white papers vary dramatically, they generally include information on: (1) the business model; (2) the technical aspects; (3) the token details, including its utility and/or rights, supply, allocation, and distribution; (4) the use of proceeds; (5) the issuing entities; (6) the law applicable to the ICO and its regulatory status; (7) the launch of the ICO—the duration, hard and soft cap, and which currencies will be accepted in exchange for tokens; and (8) the project's road map.

3.1.1 White Paper Quality and Informativeness

Empirical studies analyze the relation between quality of white papers and ICO success. Lyandres et al. (2019) find that the number of unique words in the white paper is positively associated with the amount raised and with the probability of the issued token being listing on crypto exchanges. Amsden and Schweizer (2018), Bourveau et al. (2018), and Fisch (2019) find that the length of the white paper is positively associated with the amount raised. In addition, Bourveau et al. (2018) find positive relationship between successfully completing an ICO and the informativeness of white papers.

Other studies show that investors do not trust voluntarily disclosed information in ICOs (Blaseg 2018). Ante et al. (2018) find that while the existence of white paper is important, there is no significant relationship between white paper quality and amount raised in ICOs. As the ICO market is still not as regulated compared to the IPO market, the information disclosed in the white paper is unaudited, and hence limited and sometimes misleading (Feng et al. 2018). Therefore, ICO investors may not have enough tangible reference points for their investment decisions (Momtaz 2019b; Benedetti and Kostovetsky 2018).

The remaining of this subsection summaries studies that look at disclosures of various type of information on white papers and their relations to ICO success.

3.1.2 Legal Aspects of the White Paper

On the legal side, Adhami et al. (2018) find that only in 8% of ICOs specifies the legal jurisdictions on the white papers. Zetzsche et al. (2018) find that only 28.5% of ICOs specify the applicable law and that most white papers do not provide information about the regulatory status of an ICO. As most ICOs in their samples are successful, these results suggest that potential investors are insensitive to regulatory issues. An alternative interpretation for these results is that token issuers—especially in the early days of the market—have been unable to specify the applicable law and jurisdiction due to regulatory uncertainty. Changing ICO regulations call for more recent studies over a longer sample period to examine the impacts of legal aspects and legal disclosure.

3.1.3 Technical Aspects of the White Paper

Focusing on the technical aspects: while investors are insensitive to regulatory issues, empirical evidence suggests that potential investors are very sensitive to technical aspects. Fisch (2019) finds that having a technical white paper significantly increases the amount raised. Similarly, Feng et al. (2018) find that the disclosure of blockchain architecture selected by the issuers are positively associated with the amount raised in ICOs. Lyandres et al. (2019) document that the probability of listing increases proportionately to the technical language in the white paper. Momtaz (2019a) shows that the market uncertainty derived from technical issues has a much stronger negative effect on ICO returns than regulatory actions. These results suggest that investors interpret a technical white paper as a strong indicator of issuers' underlying technological capabilities, which is an important aspect of technical blockchain projects.

3.1.4 Disclosure on the Use of Proceeds and Token Allocation

While the disclosure on the use of proceeds is required in IPOs (Leone et al. 2007), studies find that the majority of ICOs do not disclose information about the use of proceeds. For instance, Adhami et al. (2018) find only 30.8% of their sample ICOs have disclosed such information.

In addition, empirical studies have examine the relationship between the disclosure of the use of proceeds and ICO success. Howell et al. (2018) find that the use of proceeds disclosure is positively associated with token liquidity and the amount raised in ICOs. However, Bourveau et al. (2018) find that such disclosure is not related to successful ICO completion. Again, the contradictory results are mostly due to different ICO samples.

ICOs generally feature in their white paper information about token allocation—the fraction of tokens allocated to founders, advisors, early investors, and general investors, etc. Bourveau et al. (2018) find that disclosing information about token allocation is negatively associated with the amount raised. This may indicate that tokens allocation is often not optimal, and hence if issuers disclose information about it, it negatively affects the fundraising.

3.1.5 Management Team and Advisers

In the traditional market for financing, a growing body of studies documents a significant and positive relationship between management characteristics (e.g. management team legitimacy, team size, education, and reputation) and financing success and performance, suggesting management team quality as an important signal in face of uncertainty.

With higher level of uncertainties in ICOs, studies have examined the relationship between team information disclosed and ICO success. For instance, Amsden and

Schweizer (2018), Bourveau et al. (2018), and Lyandres et al. (2019) all find team size to be positively associated with the probability of token listing. An et al. (2019) find that the disclosure of founders' information on education, working experience, and social network are associated with better ICO outcomes measured by the total amount raised or the speed of fundraising. Similarly, Howell et al. (2018) suggest that entrepreneurial experience is strongly associated with ICO success. Amsden and Schweizer (2018) find that having a CEO with over 500 connections on LinkedIn is positively associated with the amount raised in an ICO. Momtaz (2019b) shows that CEO's loyalty is negatively related to ICO underpricing and positively related to firm's long-run success. Momtaz (2019c) examines the impact of CEO emotion on ICO underpricing and finds that firm experience more underpricing when CEOs signal fear or anger in photos and video materials.

Besides a management team, ICO projects usually have advisors on both business and technical sides. Studies have looked at the relationship between ICO advisors and success. Amsden and Schweizer (2018) find that having more advisors is positively associated with the amount raised and with having tradable tokens after the ICO ends. Giudici and Adhami (2019) document that advisory committee size is positively correlated to ICO success. Ante et al. (2018) find management team size, network size, and the number of advisors are positively associated with the amount funds raised in ICOs. Advisors contribute to ICO success either through the expertise and network they bring or through a signaling effects that projects that attract more advisors tend to be of high quality. However, a recent blog post from by Yavin (2018) argues that some ICO advisors join projects without conducting fundamental due diligence and therefore do not signal high quality or contribute to a project's success.

3.2 *Token Design and Tokenomics*

This subsection discusses the token design and tokenomics (token economics) issues such as token type according to the rights a token grants its holders, fraction of total generated tokens for sale, soft cap and hard cap, token supply and price, lock-up period, sales and resale restrictions, and currency accepted to purchase the token. All these details are important in ensure an ICO success.

3.2.1 Token Type

As discussed in the Introduction, two types of tokens issued through token offerings are utility tokens and security tokens. Utility tokens provide access to service or product the issuer will provide without transferring ownership or control rights of the issuing companies (Momtaz 2019a). Security tokens grant their holders financial rights, similar to equity, debt, and other financial instruments (Collomb et al. 2018). Security tokens are in most jurisdictions subject to securities regulations as their value is based on performance of the underlying assets (Hacker and Thomale 2017; Klayman 2018; Momtaz et al. 2019).

Empirical studies show that the most prevalent tokens at the time of writing are still utility tokens (e.g., Adhami et al. 2018). Examining token type and ICO success, Howell et al. (2018) find that tokens that convey utility-like rights are more likely to succeed. Fisch (2019), however, analyzes the relation between utility token and ICO success, and finds no significant difference between security tokens and utility tokens with regard to the amount raised. Nevertheless, these results should not be the determining factor on whether a blockchain venture should issue utility tokens versus security tokens, as it should depends on the overall token design and business model of the blockchain venture. On a related matter, there are projects using a dual token model that issues both utility tokens and security tokens (Damani and Gross 2018).

3.2.2 Fraction of Tokens for Sale

After the token issuer generated all the tokens, usually using a smart contract, it needs to decide the percentage of all minted tokens for sale to raise capital. Empirical studies find that the average fraction of tokens for sale in an ICO is between 54 and 61%.[1] Studies have also find that offering a higher percentage of tokens for sale is negatively related to the amount raised and the probability of taken tradability (Amsden and Schweizer 2018; Giudici and Adhami 2019; Lyandres et al. 2019). These results suggest that as ICO investors face a high degree of uncertainty because of the unregulated and opaque nature of the ICO market, a higher fraction of tokens retained by issuers signals that high quality of the project and commitment of the founding team (Davydiuk et al. 2019). In fact, Davydiuk et al. (2019) show that greater retention is related to better post-ICO performance.

3.2.3 Soft Cap and Hard Cap

Soft Cap When launching a token sale, issuers must decide whether to include a soft cap requirement. A soft cap is the minimum amount of funds an issuer aims to raise. If an ICO fails to reach the soft cap requirement, funds are usually returned to investors. Li and Mann (2018) refer it as the "all-or-nothing" clauses in ICOs and Lee et al. (2018) argue that a soft cap requirement reduces investor risk. Despite the potential benefits to investors, empirical studies find that soft cap requirements are not very common, with Amsden and Schweizer (2018) finding 32% of their sample

[1]Both Benedetti and Kostovetsky (2018) and Amsden and Schweizer (2018) find that the average percent of all tokens sold during the ICO is 60%. Fisch (2019) finds 56%. Howell et al. (2018) find that the average is 54%. Lee et al. (2018) find 57% among successful and 61% among failed ICOs. Lyandres et al. (2019) finds 57%.

specifying a soft cap requirement and Bourveau et al. (2018) finding it in only 24% of their sample.

Extant research find mixed results on the relationship between soft cap and ICO success. Amsden and Schweizer (2018) find that a soft cap requirement is positively associated with the amount raised. Lee et al. (2018) suggest that having a soft cap does not improve ICO success. Bourveau et al. (2018) argue that including a soft cap requirement is negatively associated with the amount raised.

Hard Cap is an issuer's decision on the maximum amount to raise in an ICO. Empirical studies find that the average hard cap ranges from $43 to $93 million, but the distribution is highly skewed with a median value of $20 to $23 million.[2] Studies also suggest that ICOs tend to set high hard caps that they are unlikely to reach. For instance, Lyandres et al. (2019) document that ICOs are able to raise on average 46% of their hard cap, and that only 26% of ICOs reach the hard cap; and Lee et al. (2018) find that only 12.2% of ICOs hit their hard cap. In addition, Lyandres et al. (2019) note that a higher hard cap is negatively associated with ICO success.[3] These results are consistent with findings of the IPO and crowdfunding literature that large offerings send a negative signal to the market (Lyandres et al. 2019; Mollick 2014).

3.2.4 Token Supply

Token issuers usually have a fixed token supply, expecting a token's price to increase with rising market demand for the token. Catalini and Gans (2019) theorize that in order to maximize the amount raised in an ICO, the growth rate in token supply between subsequent periods should be zero, i.e., ICOs should have a predetermined fixed token supply. Consistent with this theoretical model, Howell et al. (2018) find that the ability to create future tokens is negatively related to the amount raised. However, Cohney et al. (2018) use a sample of top 50 ICOs in 2017 and find the over 20% of ICOs that made promises of fixed token supply failed to reflect these promises in the actual smart contract code.

---

[2]Specifically, Lyandres et al. (2019) find that the mean hard cap is $93 million, while in more than 50% of the ICOs, it is larger than $20 million, highlighting the skewness. Benedetti and Kostovetsky (2018) find that the average hard cap is approximately $43 million (median = $23 million). Lee et al. (2018), find that the average hard cap for successful ICOs is approximately $88 million (median = $22 million).

[3]Specifically, they find that the ratio of amount raised normalized by hard cap is negatively associated with amount raised, with a 1% increase in hard cap associated with a 0.06–0.08% reduction in the normalized amount raised.

3.2.5 Lock-up Mechanism

An issuer must decide whether early contributors and founders would be required to commit to a lock-up period, during which they will be prevented from selling their tokens (Cohney et al. 2018). In a theoretical study, Cong et al. (2018) argue that the incentive-compatible tokens should include a lock-up period. Empirical studies find that lock-up mechanism is positively associated with ICO success. Specifically, Bourveau et al. (2018) find that ICOs with longer lock-up periods for insider raise more capital. Consistently, findings of Howell et al. (2018) suggest that token vesting information is strongly associated with secondary market liquidity and first-day trading volume. These results suggest that lock-up is a signal of quality.

Nevertheless, Cohney et al. (2018) find that in practice, many ICOs make promises regarding lock-up mechanisms but fail to reflect them in the source code. They compare the promises made in the disclosure documents with the actual functionality of the digital tokens for the top 50 ICOs that raised the most capital in 2017, and find that of the 37 ICOs that promised a lock-up mechanism, 78% did not code it.

3.2.6 Currency Accepted

A crypto token sale is usually conducted through the project's website, where investors are required to transfer money (either crypto or fiat currencies) to a smart contract address, which then transfers a pre-determined amount of tokens to the investors. Empirical studies find that on average ICOs accept two types of currencies (e.g. Amsden and Schweizer 2018; Howell et al. 2018). Lee et al. (2018) find that ICOs that accept multiple currencies are significantly more likely to succeed and have higher gross proceeds. There are in general two interpretations. First, accepting multiple cryptocurrencies requires significant technical expertise, and thus signals project quality (Amsden and Schweizer 2018). Second, multiple payment options are valuable in crypto markets, given the volatile nature of cryptocurrencies.

3.3 *Legal Aspects*

3.3.1 Sales Restrictions

As token issuances are conducted through the Internet, anyone with an Internet could have access to it (Rohr and Wright 2017). However, due to regulatory restrictions, ICOs may decide to exclude residents from certain jurisdictions. For instance, Rhue (2018) finds that 33% of her sample ICOs exclude Chinese citizens and 27% exclude US citizens.

Studies have also analyzed the relationship between jurisdiction restrictions and ICO success. Lee et al. (2018) find that ICOs that restrict sales in certain countries are less likely to succeed. Similarly, Momtaz (2019b) shows that the number of country restrictions is positively associated with ICO underpricing, suggesting that issuers that choose to reduce the set of potential investors need to offer higher incentives for the remaining. However, he also finds that ICOs that restrict countries are more likely to be successfully listed on crypto exchanges. A possible reason for this is that by preventing certain countries from participating in the ICOs, the issuer reduces the risk of regulatory actions.

For specific jurisdiction, Bourveau et al. (2018) find that ICOs that restrict US investors from participating are more likely to be successfully completed and to raise more capital. In line with Momtaz (2019b)'s interpretation, they suggest that this may reduce the risk of future SEC regulation and intervention. On the other hand, Howell et al. (2018) find that restricting US investors is unrelated to success (higher liquidity and volatility).

Studies have found that blockchain companies strategically choose their ICO locations for regulatory purposes, and therefore their issuing jurisdictions can be different from their countries of operation (Kaal 2018; Novak 2019). Specifically, Huang et al. (2018) find that ICOs are more likely to take place in countries that actively present their regulatory intentions, instead of banning ICOs or taking no action. They also note that ICOs occur more frequently in countries with developed financial markets, where information communication technology is better developed. Benedetti and Kostovetsky (2018) find that listed ICOs are more likely to take place in countries with better World Bank ranking in Rule of Law and higher GDP per capita. In addition, a report by Fabric Venture and Token Data (2018) shows a significant difference between the leading countries from a legal domicile perspective and the leading countries from founders' location perspective. For example, in 2017, legal entities located in Switzerland raised $1.06 billion compared to $177 million raised by founders from Switzerland.

3.3.2 Know Your Customer (KYC) Policies

The decentralized nature of cryptocurrencies, along with their anonymity, increases the risk of money laundering and terrorism financing (Luu 2018). Therefore, know your customer (KYC) policies are necessary for ICOs. Rhue (2018) finds that 45% of ICOs feature a KYC procedure and Lyandres et al. (2019) find that 22% of ICOs requires KYC and that 25% of ICOs feature a whitelist.

Studies that look at the relationship between adopting KYC policies and ICO success find mixed results. Lee et al. (2018) find a negative relation between the existence of KYC policies and successfully meeting ICO fundraising goals. They suggest that the KYC policies have the potential of reducing demand by investors who do not want to reveal their identity. In line with these results, Momtaz (2019a) documents a negative relation between ICO underpricing and adopting KYC. He suggests that this result is consistent with information eliciting theories in IPOs, according to which entrepreneurs get to know their potential investors during the book-building period, and can thus price their tokens more accurately. On the other hand, Lyandres et al. (2019) and Burns and Moro (2018) find that whitelist or KYC is positively related to the amount raised, suggesting that adopting KYC signals legitimacy and quality of the projects.

3.4 Technology and Source Code

As ICOs are issuance of crypto tokens through the Internet, issuers in general disclose their underlying code on an online code repository (usually the GitHub). For instance, Amsden and Schweizer (2018) find that 48% of ICOs disclosed their source code on GitHub and Adhami et al. (2018) find 40% of ICOs provided source code.

Empirical studies find that source code disclosure is positively and significantly associated with (1) successfully completing the ICO and (2) the probability of tokens being listed on crypto exchanges. For instance, Adhami et al. (2018) find that projects with full or partial code transparency counted for only 20.8% of the failed offerings, whereas those without any code made publically available are associated with 70.8% of the failures. Bourveau et al. (2018) document similar results that 51% of issuers who successfully completed an ICO have disclosed their source code, compared to only 15% of issuers who have failed. Blaseg (2018) finds that young companies that disclose higher quality source codes are more likely to list on a crypto exchange shortly after ICOs.

Adhami et al. (2018) argue that source code disclosure allows potential investors to pre-assess the technical validity of the project, and thus sends an important signal of the technical capability of the issuers. In line with these results, Rhue (2018) demonstrates that the number of bugs in the token code, identified by Etherscan, is negatively and significantly associated with market cap. In addition, Howell et al. (2018) find a negative relation between days from last token revision and token liquidity, which suggests that being active on GitHub is a positive signal for potential investors.

Information asymmetry is particularly severe with regard to the technical aspects of ICOs. While investors tend to be highly sensitive to the technical aspects, and specifically to source code disclosure, they are insensitive to the quality of the code. Cohney et al. (2018) support this argument empirically by showing significant mismatches between promises made in white papers and the actual code, and that the number of uncoded promises does not affect the amount raised. In the long term, however, Cohney et al. (2018) find a negative correlation between the number of uncoded promises and price appreciation, suggesting that information asymmetry decreases as time goes by.

3.5 Marketing and Social Media

3.5.1 Social Media

Unlike IPOs, in which the underwriter is responsible for marketing the venture to potential investors, in the case of ICOs, the marketing process is carried out by token issuers through social media (Rhue 2018). Social media serve as important marketing and communication channels for announcing ICOs and distributing information about the underlying tokens and project development progress (Chanson et al. 2018). As a result, social media reduces the information asymmetry and uncertainty around the project.

Empirical studies show that most ICOs are active on social media platforms, with the most common being Telegram and Twitter. Specifically, Rhue (2018) finds that ICOs usually engage a median of eight social media platforms, such as Twitter, Facebook, Telegram. Howell et al. (2018) show that in a sample of 453 ICOs, 83% have a Telegram group with an average of over 5000 members and that 97% have an official Twitter account with average of 22,200 followers.

Research has found that social media presence and activities are among the major factors that influence ICO success. Bourveau et al. (2018) find that social media activity is positively associated with successfully completing an ICO, the amount raised, and the liquidity of the token issued. Authors such as Benedetti and Kostovetsky (2018), Fisch (2019), Burns and Moro (2018) have all documented a positive relationship between Twitter activity and ICO success. Amsden and Schweizer (2018) show that having a Telegram group is positively correlated with the probability of having tradable tokens after ICO completion. In addition, Howell et al. (2018) find that the number of followers on Twitter and Telegram is positively associated with liquidity, but only the former is significantly correlated with long-term returns.

These results suggest that as social media platforms play a vital role in the ICO sphere, token issuers and blockchain-based ventures may strategically use social media to influence investors' behavior. Momtaz et al. (2019) find that market sentiments and market liquidity are strongly associated with listing and with social media activity, suggesting that issuers have an incentive to create a positive investor sentiment and that they can do it through social media.

Other studies find that ventures tend to open social media accounts just in time for the token sale, and that social media activity drops following ICO completion. Specifically, Lyandres et al. (2019) find that social media activities on Medium, Twitter and Reddit significantly decrease following ICO completion. Benedetti and Kostovetsky (2018) show that the average Twitter account age is about 8 months with a median of only 3 months, suggesting that a large number of ICOs open social media accounts just in time for their ICO. These results imply that blockchain issuers are aware of social media's effects and strategically use them to generate hype during the token sale.

Lyandres et al. (2019) find that investors are, at least partially, aware of those strategies, showing that reduction in social media activity following ICO completion is significantly associated with listing and with the amount raised. Similarly, Rhue (2018) finds that while an ICO has a median of eight social media links, a higher number of social media links for the project is associated with lower ROI. Similarly, Ante et al. (2018) find that Twitter and Facebook has a very small positive impact in ICO fund raised, with Bitcointalk and Reddit yielding insignificant and negative results. A possible interpretation for this result is that strategic use of social media may reduce a project's credibility.

Other studies identify investors' irrational behavior in the context of social media. Benedetti and Kostovetsky (2018), for example, document a bias for good news and overreaction to company announcements. They find that daily market return is positively correlated with today's company announcements—suggesting that "no news is bad news"—and that today's return is negatively correlated with a high level of Twitter activity from the prior month, which suggests that investors overreact to company announcements.

Along with social media activities, during the ICO campaign, token issuers generally execute bounty programs and airdrops. First, in bounty programs, issuers offer tokens in exchange for performing certain tasks. For example, some issuers may reward token rating websites for writing an article about the ICO, or individuals for translating their documents into different languages or fixing bugs in the underlying code (Glier 2018). Second, in an airdrop, issuers give free tokens to those who follow their social media accounts , in order to raise awareness of the project and encourage the token's adoption (Dale 2018; Momtaz et al. 2019).

3.5.2 ICO Ratings and Online Forums

ICOs also rely on third parties such as the ICOBench and the online discussion forums to provide information and evaluation of the projects. Discussions from these websites and forums are considered as independent analysts' opinion. Therefore, token investors who tend to follow early investors can substitute traditional underwriters' intermediary roles in ICOs and reduce information asymmetry.

Some studies show that analysts' rating from independent and unofficial websites strongly and reliably predict ICO success. For instance, Lee et al. (2018) find that projects with a higher ICOBench rating tend to have higher likelihood of successful fundraising, a quick sale, and a higher 3-month token returns.[4] Bourveau et al. (2018) show similar results when analyzing the rating scores from ICOBench and ICORating.[5] These findings suggest that even in unregulated markets, information

[4]Specifically, Lee et al. (2018) find that gross proceeds increase by $4.7 million when the average analyst rating increases by one point and that successful ICOs on average had a rating of 3.3 (out of 5), 0.7 points higher than that for failed token sales.

[5]Their results are (1) completed ICOs tend to have significantly higher ratings than failed ICOs; (2) rating is positively associated with the likelihood of completing an ICO; (3) higher ratings are strongly negatively associated with two measures of crash risk, extreme negative returns and negative return skewness; and (4) are negatively associated with post-ICO illiquidity and return volatility.

intermediaries naturally emerge, and that ICO market participants find their assessments credible. In addition, Rhue (2018) finds that ICO Drops' reputation and hype scores are positively and significantly associated with higher ROI, and that ICO reputation scores from Etherscan predict higher market cap. Similarly, Momtaz (2019b) finds that the quality of the management team, as measured by ICOBench, is positively and significantly associated with market performance and higher gross proceeds.

However, there are several major concerns regarding these independent analysts. A first issue is the inconsistency across various rating sources and websites (Rhue 2018). Second, while independent rating websites have the potential to overcome information asymmetry, some crypto bloggers have been shown to simply sell the rating scores (Hartmann 2018; Poutintsev 2018). Third, the inaccuracy of the rating can also be a concern. For instance, using variables provided by ICOBench, Bourveau et al. (2018) find a strong relationship between white paper informativeness and ICO success. However, when the authors manually analyze the association between disclosure practices (ICO team information, token allocation information, founder tokens vesting period, use of proceeds, whitepaper opacity, and whitepaper length) and ICO success, they find no significant and even negative association. Fourth, while independent rating sites have been found to be strongly and reliably associated with ICO success, Cohney et al. (2018) find that only one of the top five rating sites by Alexa ranking post source code information. Again, these results highlight the challenges faced by investors in finding reliable information. The discussions above show that independent analysts' ratings, which may reduce this asymmetry, are inaccurate at best.

Regarding online discussion forums, Mai et al. (2018) find that Internet forum have a stronger impact on future Bitcoin value than Tweeter. Similarly, Chanson et al. (2018) find a significant relation between the number of threads where an ICO is mentioned on selected online discussion forums in the 30 days prior and underpricing. These results suggest that media-provided content has a strong influence on crypto investors' behaviors.

4 ICO Sale

4.1 *ICO Presale*

ICOs generally conduct private and public presales (together, the "presale") prior to the full ICO launch, which target mainly institutional investors and VCs, and offer them discounts or bonuses in exchange for taking more risk (investing in an early stage). Empirical evidence shows that presales are a common practice. Specifically, the percentage of presale is found to be 40% in the ICO samples of Benedetti and Kostovetsky (2018), 33% in Adhami et al. (2018), 64% in Fisch (2019), and 44% in Momtaz (2019a). In addition, Benedetti and Kostovetsky (2018) find that presales have become more popular over time, with an average incidence of 1% for ICOs completed before July 1, 2017, 29% for the second half of 2017 ICOs, and 57% for 2018 ICOs.

The purpose of the presale is twofold: (1) to finance the costs of promoting the ICO in the early stage and (2) to provide an indication of the demand for the token, thus helping to determine an appropriate price for the launch, similar to the book-building process in IPOs (Howell et al. 2018). The presale targets larger investors with a minimum contribution threshold, and offers them discounts or bonuses in exchange for taking more risk by investing at an early stage (Howell et al. 2018).

Studies have found that presale is positively associated with ICO success. For instance, Lyandres et al. (2019) find that the amount raised increases strongly in the presence of a presale and that ICOs with a successful presale are more likely to be listed. Lee et al. (2018) show that 39.5% of successful ICOs included a presale, compared to 21.3% of failed ICOs and that including a presale can boost the success likelihood by 15.2 percentage points. Li and Mann (2018)'s model suggest that investors are heterogeneously informed and that investors with a relatively high signal would join early and those with a relatively week signal would follow the crowd. To put it another way, presales are interpreted by later investors as evidence that earlier investors held favorable information, and thus trigger an information cascade. This interpretation is consistent with both the crowdfunding and IPO literature (Lyandres et al. 2019).

Other studies, however, find that presales to be negatively associated with ICO success or have no significant relation. For instance, Momtaz (2019a) finds that a presale reduces the total funding amount raised in the actual ICO by an average of $7.11 million. A possible explanation, as Amsden and Schweizer (2018) suggest, is that presales may indicate that a firm is unsecure about the ICO. Another explanation is that to attract sophisticated investors, firms need to offer high bonuses during presales and high bonuses may lead to pump-and-dump, as well as Ponzi schemes (Li and Mann 2018; Amsden and Schweizer 2018). Empirical evidence suggests that investors are aware of these risks, so that offering bonuses, particularly higher ones, predicts failure and lower first-day returns on the secondary market. Specifically, Lee et al. (2018) find that ICOs offering large bonuses are 10.9% less likely to succeed, and that high bonus is negatively related to the amount raised in an ICO, the first-day sales volumes, and first-day return.

4.2 ICO Sale

Empirical studies find that on average ICOs raise around $15 million to 20 million. Specifically, Momtaz (2019a) finds that the average (median) amount raised in an ICO is $15.1 ($5.8) million. Lyandres et al. (2019) find that the average (median) amount raised is $15 million ($4 million). Amsden and Schweizer (2018) find $15.26 million ($6.04 million). In a more recent study, Fisch (2019) finds the average (median) to be $19.6 million ($11.8 million).

Studies in general find that the average duration of an ICO sale is around 25–40 days (Adhami et al. 2018; Benedetti and Kostovetsky 2018; Fisch 2019; Howell et al. 2018).[6] Empirical studies suggest that the duration of an ICO is negatively related to success. Lee et al. (2018) find that successful ICOs took an average of 30 days to complete compared to 37.8 days for failed fundraisers. Momtaz (2019a) and Fisch (2019) show that the duration of an ICO is negatively associated with the amount raised. These results are in line with the crowdfunding literature (Mollick 2014) that suggest shorter durations may encourage prospective investors to act fast.

An interesting feature of ICOs is that investors invest substantial amounts of wealth without the right to claim a fair return on their investment later on. The lack of investor protection is due to at least three reasons: ICOs are unregulated, virtual currency projects have no corporate governance, and ICOs take place without underwriting. These three reasons distinguish ICOs from conventional IPOs (Liu 2019). The IPO literature sees underwriters mainly as institutions that reduce informational asymmetries and adverse selection problems (Beatty and Ritter 1986; Benveniste and Spindt 1989; Habib and Ljungqvist 2001; Ljungqvist 2007). In the absence of underwriters, asymmetric information and adverse selection problems become key challenges in the ICO market.

So why then are investors attracted to ICOs? A potential explanation is that serious (i.e., non-fraudulent) projects have a strong incentive to reward investors in the short run by means of ICO underpricing to generate market liquidity. ICOs are unique in that issued tokens are essentially currencies that are of value for a specific platform, and that the amount of token supply is usually fixed. Therefore, the higher the demand on the platform with fixed supply of tokens, the higher the token price. Further, Trimborn et al. (2018) show that token demand is restricted by liquidity in the secondary market. ICO projects thus have an incentive to underprice their tokens in the ICO to generate market liquidity as a knock-on effect to signal platform growth prospects.

4.3 Post-ICO

After the ICO, the issuers have raised money to develop its business and technology, usually the blockchain platform. Issuers then list their tokens on crypto exchanges, and the issued tokens are traded on the secondary market.

A portion of the tokens received during the main token sale is usually reserved for founders, employees, and platform development, and/or for incentivizing future

---

[6]Specifically, Adhami et al. (2018) find that the average duration is 27 days, but that it is heterogeneous—"some ICOs close in a few days, whereas other are open for some months". Benedetti and Kostovetsky (2018) find that the average ICO lasts 37 days (median = 31). They also find that this figure has recently been rising with an average of 41 days for 2018 ICOs, Fisch (2019) finds that the average duration is 25 days. Howell et al. (2018) find that the average duration of an ICO is 40 days.

network contributors (Howell et al. 2018). These tokens are generally locked in smart contracts for a specific period or until certain development milestones have been achieved. After the lock-up period, tokenholders start to vest their tokens.

4.3.1 Exchange Listing

After the token sale ends, ICOs generally list their tokens in crypto exchanges, and the issued tokens are then traded on the secondary market. Listing is an important indicator of ICO success, as it provides the main source of token liquidity (Amsden and Schweizer 2018; Momtaz 2019a). Therefore, studies such as Amsden and Schweizer (2018) and Lyandres et al. (2019) use exchange listing as a proxy for ICO success.

Unlike IPOs, tokens do not list on a cryptocurrency exchange immediately after the ICOs. It usually takes about 1 week to 6 months for a token to be listed on an exchange, if it gets listed at all (Feng et al. 2018). Empirical studies find that the time from ICO completion to exchange listing is highly skewed, with some ICOs being listed during the token sales and other over a year after ICO completion. The average time ranges from 18.5 to 93 days.[7]

Lyandres et al. (2019) find that a token is traded on average on six different exchanges, and that the number of exchanges is positively associated with success. This suggests that exchanges are willing to trade tokens of successful ICOs and that successful ICOs are willing to pay listing fees. They also find that larger ICOs are more likely to be listed. Benedetti (2019) studies token cross-listings and find significant trading volume, liquidity, and return increase around cross-listings.

5 Conclusion

This chapter reviews empirical studies on the characteristics of initial coin offerings (ICO) and determinants of ICO success. This paper contributes to the literature by providing a discussion on all key elements in a full-cycle ICO and discuss comprehensive literature review the common practice and key success factors. Findings of this paper provide important managerial and policy implications. Regulators should identify the specific market frictions discussed in this paper and the main sources of investors irrationally and accordingly, provide a regulation framework that will protect investors and promote market efficiency. The optimal regulatory framework should address information asymmetry by using disclosure provisions. In addition, it should impose legal obligations on analysts reviewing ICOs and the marketing materials.

[7]Specifically, Lee et al. (2018) find that the average time from ICO completion to listing is 18.5 days, whereas Benedetti and Kostovetsky (2018) find that the average (median) time to be 31 (16) days. They also find that some ICOs were listed prior to the end of the ICO. Momtaz (2019a) documents an average (median) time of 93 (42) days.

References

Adhami S, Giudici G, Martinazzi S (2018) Why do businesses go crypto? An empirical analysis of initial coin offerings. J Econ Bus 100:64–75

Amsden R, Schweizer D (2018) Are blockchain crowdsales the new "gold rush"? Success determinants of initial coin offerings. https://papers.ssrn.com/sol3/papers.cfm?abstract_id=3163849

An J, Duan T, Hou W, Xu X (2019) Initial coin offerings and entrepreneurial finance: the role of founders' characteristics. J Altern Invest 21(4):26–40

Ante L, Sandner P, Fiedler I (2018) Blockchain-based ICOs: pure hype or the dawn of a new era of startup financing? J Risk Financ Manage 11(4):80

Barsan IM (2017) Legal challenges of initial coin offerings (ICO). Revue Trimestrielle de Droit Financier (RTDF) 3:54–65

Beatty RP, Ritter JR (1986) Investment banking, reputation, and the underpricing of initial public offerings. J Financ Econ 15(1–2):213–232

Benedetti H (2019) The economics of digital token cross-listings. https://papers.ssrn.com/sol3/papers.cfm?abstract_id=3267392

Benedetti H, Kostovetsky L (2018) Digital tulips? Returns to investors in initial coin offerings. Returns to Investors in Initial Coin Offerings (May 20, 2018)

Benveniste LM, Spindt PA (1989) How investment bankers determine the offer price and allocation of new issues. J Financ Econ 24(2):343–361

Blaseg D (2018) Dynamics of voluntary disclosure in the unregulated market for initial coin offerings. https://papers.ssrn.com/sol3/papers.cfm?abstract_id=3207641

Boreiko D, Sahdev NK (2018) To ICO or not to ICO: empirical analysis of initial coin offerings and token sales. https://papers.ssrn.com/sol3/Papers.cfm?abstract_id=3209180

Bourveau T, De George ET, Ellahie A, Macciocchi D (2018) Initial coin offerings: early evidence on the role of disclosure in the unregulated crypto market. https://www.marshall.usc.edu/sites/default/files/2019-03/thomas_bourveau_icos.pdf

Burns L, Moro A (2018) What makes an ICO successful? An investigation of the role of ICO characteristics, team quality and market sentiment, September 27, 2018

Catalini C, Gans JS (2019) Initial coin offerings and the value of crypto tokens. https://www.nber.org/papers/w24418

Chanson M, Gjoen J, Risius M, Wortmann F (2018) Initial coin offerings (ICOs): the role of social media for organizational legitimacy and underpricing. https://www.alexandria.unisg.ch/255399/

Chod J, Lyandres E (2018) A theory of ICOs: diversification, agency, and information asymmetry. https://papers.ssrn.com/sol3/papers.cfm?abstract_id=3159528

Clements R (2018) Assessing the evolution of cryptocurrency: demand factors, latent value, and regulatory developments. Mich Bus Entrep L Rev 8:73

Cohney S, Hoffman DA, Sklaroff J, Wishnick DA (2018) Coin-operated capitalism. https://lowellmilkeninstitute.law.ucla.edu/wp-content/uploads/2019/02/Coin-operated-Capitalism.pdf

CoinSchedule (2018) Cryptocurrency ICO Stats 2018. www.coinschedule.com/stats.html?year=2018

Collomb A, De Filippi P, Sok K (2018) From IPOs to ICOs: the impact of blockchain technology on financial regulation. https://papers.ssrn.com/sol3/papers.cfm?abstract_id=3185347

Cong LW, Li Y, Wang N (2018) Tokenomics: dynamic adoption and valuation. https://papers.ssrn.com/sol3/papers.cfm?abstract_id=3222802

Dale B (2018) So long ICOs, hello airdrops: the free token giveaway craze is here. https://www.coindesk.com/long-icos-hello-airdrops-free-token-giveaway-craze

Damani S, Gross J (2018) A dual token structure of utility and security tokens—leveraging the best of both token words. https://www.minthealth.io/a-dual-token-structure-of-utility-and-security-tokens-leveraging-the-best-of-both-token-worlds/

Davydiuk T, Gupta D, Rosen S (2019) De-crypto-ing signals in initial coin offerings: evidence of rational token retention. https://papers.ssrn.com/sol3/papers.cfm?abstract_id=3286835

Dittmar RF, Wu DA (2019) Initial coin offerings hyped and dehyped: an empirical examination. https://papers.ssrn.com/sol3/papers.cfm?abstract_id=3334166
EY (2017) EY research: initial coin offerings (ICOs). https://www.ey.com/Publication/vwLUAssets/ey-research-initial-coin-offerings-icos/$File/ey-research-initial-coin-offerings-icos.pdf
Fabric Ventures & Token Data (2018) The state of the token market: a year in review an outlook for 2018. https://static1.squarespace.com/static/5a19eca6c027d8615635f801/t/5a73697bc8302551711523ca/1517513088503/The+State+of+the+Token+Market+Final2.pdf
Feng C, Li N, Lu B, Wong MH, Zhang M (2018) Initial coin offerings, blockchain technology, and voluntary disclosures. https://papers.ssrn.com/sol3/papers.cfm?abstract_id=3256289
Fisch C (2019) Initial coin offerings (ICOs) to finance new ventures: an exploratory study. J Bus Ventur 34(1):1–22
Giudici G, Adhami S (2019) The impact of governance signals on ICO fundraising success. J Ind Bus Econ 1–30
Glier G (2018) What are ICO bounty programs? https://blocktoken.ai/what-are-ico-bounty-programs
Habib MA, Ljungqvist AP (2001) Underpricing and entrepreneurial wealth losses in IPOs: theory and evidence. Rev Financ Stud 14(2):433–458
Hacker Moon (2017) Early investors are making 50,000% returns on ICOs. https://hackernoon.com/investors-are-making-50-000-returns-on-icos-32432bc741d1
Hacker P, Thomale C (2017) Crypto-securities regulation: ICOs. Token sales and cryptocurrencies under EU Financial Law, November 22, 2017
Hartmann M (2018) This is how easy it is to buy ICO ratings—an investigation. https://medium.com/alethena/this-is-how-easy-it-is-to-buy-ico-ratings-an-investigation-13d07e987394
Howell ST, Niessner M, Yermack D (2018) Initial coin offerings: financing growth with cryptocurrency token sales. https://www.nber.org/papers/w24774
Huang W, Meoli M, Vismara S (2018) The geography of initial coin offerings. Small Bus Econ 1–26
Kaal WA (2018) Initial coin offerings: the top 25 jurisdictions and their comparative regulatory responses. https://medium.com/semadaresearch/initial-coin-offerings-the-top-25-jurisdictions-and-their-comparative-regulatory-responses-4b8c9ae7e8e8
Klayman JA (2018) Mutually assured disruption: the rise of the security token. Blockchain & Cryptocurrency Regulation
Krypital Group (2018) Security token case analysis: Aspen coin – the first real estate security token offering. https://medium.com/krypital/security-token-case-analysis-aspen-coin-the-first-real-estate-security-token-offering-bbbcc52ace5
Lee J, Parlour CA (2019) Consumers as financiers: crowdfunding, initial coin offerings and consumer surplus. https://www.chapman.edu/research/institutes-and-centers/economic-science-institute/_files/ifree-papers-and-photos/parlour-lee-consumers-as-financiers-2019.pdf
Lee J, Li T, Shin D (2018) The wisdom of crowds and information cascades in FinTech: evidence from initial coin offerings. https://papers.ssrn.com/sol3/papers.cfm?abstract_id=3195877
Leone AJ, Rock S, Willenborg M (2007) Disclosure of intended use of proceeds and underpricing in initial public offerings. J Account Res 45(1):111–153
Li J, Mann W (2018) Initial coin offering and platform building. https://papers.ssrn.com/sol3/papers.cfm?abstract_id=3088726
Liu C (2019) FinTech and its disruption to financial institutions. In: Organizational transformation and managing innovation in the fourth industrial revolution. IGI Global, Hershey, pp 104–124
Liu C, Wang H (2019) Crypto tokens and token offerings: an introduction. In: S. Goutte, K. Guesmi, S. Saadi (eds), Cryptofinance and mechanism of exchange: The making of virtual currency. Forthcoming.
Ljungqvist A (2007) IPO underpricing. In: Espen Eckbo B (ed) Handbook of empirical corporate finance. Elsevier, Amsterdam, pp 375–422

Luu L (2018) With blockchain, knowing your customer is more important than ever. https://www.forbes.com/sites/luuloi/2018/05/17/with-blockchain-knowing-your-customer-is-more-important-than-ever/#1bf0b509559c

Lyandres E, Palazzo B, Rabetti D (2019) Are tokens securities? An anatomy of initial coin offerings. http://abfer.org/media/abfer-events-2019/annual-conference/corporate-finance/30-P1-Are-Tokens-Securities.pdf

Mai F, Shan Z, Bai Q, Wang X, Chiang RH (2018) How does social media impact Bitcoin value? A test of the silent majority hypothesis. J Manage Inf Syst 35(1):19–52

Mollick E (2014) The dynamics of crowdfunding: an exploratory study. J Bus Ventur 29(1):1–16

Momtaz P (2019a) Initial coin offerings. https://papers.ssrn.com/sol3/papers.cfm?abstract_id=3166709

Momtaz P (2019b) Initial coin offerings, asymmetric information, and loyal CEOs. https://papers.ssrn.com/sol3/papers.cfm?abstract_id=3167061

Momtaz P (2019c) CEO emotions and underpricing in initial coin offerings. https://papers.ssrn.com/sol3/papers.cfm?abstract_id=3305765

Momtaz PP, Rennertseder K, Schröder H (2019) Token offerings: a revolution in corporate finance? https://papers.ssrn.com/sol3/papers.cfm?abstract_id=3346964

Novak M (2019) Crypto-friendliness: understanding blockchain public policy. Journal of Entrepreneurship and Public Policy

Ofir M, Sadeh I (2019) ICO vs IPO: empirical findings, market frictions and the appropriate regulatory framework. Market Frictions and the Appropriate Regulatory Framework, February 19, 2019

Poutintsev F (2018) Beware of ICO Bench. https://hackernoon.com/beware-of-ico-bench-97addacfedc7

Rhue L (2018) Trust is all you need: an empirical exploration of initial coin offerings (ICOs) and ICO reputation scores. https://papers.ssrn.com/sol3/papers.cfm?abstract_id=3179723

Rohr J, Wright A (2017) Blockchain-based token sales, initial coin offerings, and the democratization of public capital markets. Cardozo Legal Studies Research Paper, 527

Trimborn S, Li M, Härdle WK (2018) Investing with cryptocurrencies – a liquidity constrained investment approach. https://papers.ssrn.com/sol3/papers.cfm?abstract_id=2999782

Yavin O (2018) Enough is enough! – corrupt ICO advisors, experts, and listing sites. https://www.cointelligence.com/content/ico-expert-corruption/

Zetzsche DA, Buckley RP, Arner DW, Föhr L (2018) The ICO gold rush: it's a scam, it's a bubble, it's a super challenge for regulators. Harv Int Law J 63(2)

Initial Coin Offerings (ICOs): Risks, Regulation, and Accountability

Usman W. Chohan

Abstract The emergence of the cryptocurrency as an investment vehicle has brought the phenomenon of Initial Coin Offerings (ICOs) into the spotlight, since they provide rapid access to capital for new ventures, but suffer from drawbacks relating to regulation and accountability. In that regard, this chapter provides a review of the recent literature on ICOs before proceeding with a discussion of the regulatory and other risks that ICOs pose for market participants, thereby encouraging a broader discussion about where such a novel capital-raising mechanism may lie in the investment universe, and how the weaknesses of ICOs may be addressed so as to better leverage its strengths towards value creation and innovation.

1 Introduction

The ascent of cryptocurrencies as both investment vehicle and cultural phenomenon (Conley 2017; Chohan 2017a) has led to the flurry of research and investor interest in the field (Lee et al. 2018; Chohan 2017c, 2017e); and while the number of cryptocurrencies has grown tremendously in the past few years, most of them have not, even for so short an interval, stood the test of time. Worse still, numerous cryptocurrencies have been launched as opportunistic pretexts for theft, Ponzi schemes, fraudulent practices, and commercial deceit (see Benedetti and Kostovetsky 2018; Venegas 2017; Chohan 2018a–c, 2019a, b). The losses incurred have been significant, as shall be discussed later in this chapter—but nevertheless, as a result of considerable monetary damage to a non-specialist general public, there is now a widespread call for greater regulatory accountability and oversight of the cryptocurrency space (GAO 2014; Clayton 2017; Chohan 2018c).

At the heart of the commercial process for dealing with cryptocurrencies is the Initial Coin Offering ("ICO" see Howell et al. 2018; Li and Mann 2018), which is somewhat (but not entirely) analogous to the Initial Public Offering (IPO) that is the bedrock for large-scale commercial ownership and participation in capitalism. An

U. W. Chohan (✉)
School of Business, UNSW Canberra, Canberra, Australia
e-mail: u.chohan@adfa.edu.au

S. Goutte et al. (eds.), *Cryptofinance and Mechanisms of Exchange*, Contributions to Management Science, https://doi.org/10.1007/978-3-030-30738-7_10

ICO, also termed token sale or crowdsale, is the mechanism by which capital is raised from investors through the emission of cryptocurrency monetary units of "coins" (or "tokens," see Adhami et al. 2018; Chohan 2017d), usually (but not necessarily) in exchange for traditional units of currency such as the United States dollar, the Yen, or the Euro (Fisch 2019; Chohan 2019a, b), often expressed as a percentage of total newly issued currency (Catalini and Gans 2018). ICOs may sell either cryptocurrency, or may sell a right of ownership or royalties to a project, and this is what contrasts them with IPOs, which sell a share in the ownership of a company itself (Li and Mann 2018; Chohan 2018a–c).

Adhami et al., describe ICOs as "open calls for funding promoted by organizations, companies, and entrepreneurs to raise money through cryptocurrencies, in exchange for a "token" that can be sold on the Internet or used in the future to obtain products or services and, at times, profits," (2018, p. 64). In essence, ICOs are a new motor for raising investment capital (Howell et al. 2018; Lee et al. 2018; Adhami et al. 2018; Catalini and Gans 2018), and they offer "significant promise for new startups in the cryptocurrency space as means of quicker and easier capital raise," (Chohan 2017d, p. 3). There are at least three conceivable advantages of using ICOs: (1) reducing the cost of raising capital, (2) positive network effects with a built-in customer base (see also Benedetti and Kostovetsky 2018), and (3) a secondary trading market in issued tokens (see Adhami et al. 2018, p. 64).

However, ICOs have mostly occurred in the online realm that lies beyond regularized and traditional finance, devoid of the structures of financial regulation which allow for capitalism to function in a more stable and lawful manner (Fisch 2019; Howell et al. 2018; Chohan 2017b, 2019a, b). ICOs are "bypassing any regulation that normally applies to businesses placing securities to retail investors, [and so] dozens of developer teams and entrepreneurs collect money in absence of official prospectuses, with no particular protection for contributors and disclosing only a very limited set of information," (Adhami et al. 2018, p. 65). Furthermore, "these ventures often resemble the startups that conventionally finance themselves with angel or venture capital (VC) investment, though there are many scams, jokes, and tokens that have nothing to do with a new product or business," (Howell et al. 2018, p. 1).

Instead, ICOs (and cryptocurrencies more fundamentally) lie philosophically within cryptoanarchist thought, which seeks to cultivate decentralized, autonomous, and voluntary exchange among individuals in a manner that protects their identities and therefore their risk of persecution by structures of authority (Chohan 2017f). Laudable as those ideals may be for some, cryptoanarchist principles assume a very high degree of trust, or to put it more correctly, the lack of a need for trust ("trustlessness", see Chohan 2019c) among participants. The massive frauds of the ICO space over the last few years, however, have put the praxis of utopian cryptoanarchist ideals into serious question. More specifically, ICO non-accountability and non-oversight have raised a public furore which is now being met, however haphazardly (see comparisons in Chohan 2017b), by traditional regulatory authorities.

The purpose of this chapter, then, is to discuss the need for ICO regulation and accountability (see initial work in Chohan 2017d). It does so by first providing a review of the recent literature on ICOs before considering the regulatory and other risks that ICOs pose (and have already posed) for market participants. The chapter then notes the uncoordinated and divergent international regulatory responses to ICOs, before highlighting areas of further research into ICO accountability and regulation. In that regard, this chapter should be seen as a call towards a broader discussion about where such a novel capital raising mechanism may lie in the investment universe, and how the weaknesses of ICOs may be addressed so as to better leverage its strengths towards value creation and innovation.

2 Academic Interest in ICOs

Unlike the volatile prices of cryptocurrencies, the academic literature on cryptocurrencies has risen in a more steady and graduated manner (see review in Chohan 2019c). However, although the scientific and mathematically-oriented literature on cryptocurrencies has risen more broadly, particularly in the form of white papers delineating variants of coins that address theoretical or practitioner problems (see discussion in Fisch 2019, pp. 9–12), the social (including the accountability) dimension of crypto-instruments more broadly and ICOs specifically has remained somewhat unexplored despite calls for greater policy and research engagement (see Venegas 2017; Kaal and Dell'Erba 2017; Chohan 2017d, 2018a–c). This brief section therefore reviews the salient observations on the literature specific to ICOs.

Chohan's working papers on cryptocurrency accountability (2017a–f, 2018a–c, 2019a–c) draw upon the financial accountability literature to raise the most explicit call-to-arms for improving the oversight and accountability of the cryptocurrency space, both through national and international initiatives. This list includes the first paper (Chohan 2017d) to highlight the "risks, regulation, and accountability" of ICOs specifically. The common thread among these papers has been that cryptocurrencies and their ICOs offer the promise of innovation but also pose a threat in the absence of accountability mechanisms. As such, the existing international financial structure has been caught off-guard and has only recently begun to tackle issues of cryptocurrency oversight, regulation, and enforcement (Chohan 2017a–f), and that too in a reactive manner.

In a similar vein, Fisch has noted that "ICOs are characterized by a considerable amount of information asymmetry, for example, because ventures are typically in early stages [and] the amount of objective information surrounding ICOs is very low, and there is thus considerable potential for fraud," (2019, p. 5). Kaal and Dell'Erba stress that ICOs are inherently early-stage investments and contain the concomitant risks of early lifecycle investments in any case (2017), and Fisch emphasizes that "formal disclosure requirements in ICOs are largely absent," (2019, p. 10). However, Catalini and Gans note that "even in the absence of fraud and incompetence, how precisely tokens have value in the absence of additional rights on the venture is not obvious," (2018, p. 3).

What is the purpose of ICOs in absolute terms and relative to traditional financial structures? Catalini and Gans identify the subjacent logic of strong ICO demand to be that they "allow entrepreneurs to generate buyer competition for the token, which, in turn, reveals consumer value without the entrepreneurs having to know, ex ante, consumer willingness to pay," (2018, p. 1), and this in turn "may increase entrepreneurial returns beyond what can be achieved through traditional equity financing." It is also important to note that there is a "utility" aspect to certain ICOs, in that the capital raised allows users exclusive access to services through purchase and trading of a specific coin (Conley 2017). Those rights of access serve as a utility, and this utilitarian approach has been (somewhat incorrectly) justified as a basis for having ICOs avoid the regulation-regime of securities (Clayton 2017).

Adhami et al. suggest three conceivable advantages of ICOs as a capital-raising vehicle (2018): (1) in cost-reductions in capital raising, by avoiding intermediaries and payment agents (see also Howell et al. 2018); (2) in a built-in customer base and positive network effects through platform development; and (3) in the creation of a secondary market through trading of the tokens themselves (Adhami et al. 2018). For the former point, Catalini and Gans note that ICOs rely upon "blockchain technology lowering both the cost of verification of transaction attributes—which allows for self-custody of digital assets—and the cost of coordinating economic activity over the internet," (2018, p. 2).

What raises risks from an oversight perspective (Chohan 2019a–c), but creates an opportunity from a pricing perspective, is that "conditional on successfully raising enough funds to cover development costs, the value of an ICO is independent of the anticipated growth of the platform," (Catalini and Gans 2018), or in Fisch's words, they "do not seem to relate to the venture's underlying capabilities and are highly specific to the ICO context," (2019, p. 2). Adhami et al. (2018) also found that the success rate of the ICOs was initially quite high (the tenor has changed since their publication; as has the regulatory environment).

But does this mean that most conceivably viable ventures would do well to ride the wave of ICOs? Catalini and Gans dismiss this and note that "a viable venture, which could have successfully raised capital through traditional sources, may fail to raise enough funds to cover its costs through an ICO," and particularly so when "the venture is long-lived, and is consistent with the rise of hybrid arrangements where ventures raise a traditional venture capital round before issuing tokens to the public or to accredited investors," (2018, p. 4). Similarly, Fisch observes that "due to their highly technological nature, ICOs are not applicable to every venture. Rather, they only appeal to ventures utilizing [the distributed ledger technology that underlies blockchains], which is a narrow segment of high-tech firms," (2019, p. 7).

Adhami et al. examine 253 ICO campaigns to identify the factors that would lead to a successful ICO campaign (2018). They find that the probability of an ICO's success is higher if the code source is available, if a token "presale" is arranged, and if tokens allow contributors to access a specific service (or have a share in profits). Separate work by Fisch examining 423 ICOs using a Signalling Theory approach corroborates these findings (2019); specifically that technically-robust white papers and code source availability are important determinants in successful ICOs. Work by Howell et al. studies 453 tokens and gauges success in terms of the liquidity of tokens 6 months post-release, with generally similar findings regarding the success of ICO ventures (2018).

In contrasting the successful launches of ICOs, Adhami et al. also delineate just what "failure" for an ICO can mean, which can include factors such as: (1) not meeting a minimum funding goal, in which case the ICO should refund the proceeds to investors; (2) a hack and security flaw (see also Chohan 2018a–c); and that (3) "an ICO may reveal itself to be a scam or at least perceived as a scam by the online community, resulting in a very low or zero amount of funding," (2018, p. 67).

As Conley's discussion on the valuation of cryptocurrencies indicates, a regulatory gap in ICOs "makes it uncertain what guarantees and enforceable promises [cryptocurrency] founders make to token holders. A white paper is not a contract!" (2017, p. 22, emphasis in the original). For internal governance, Conley also remarks that investors ("token holders") are "sometimes given collective control over a variety of aspects of a project, but almost never full control or proportional sharing of profits," and this is troublesome because "when any aspect of control is separated from profit sharing, serious incentive problems are created," (2017, p. 23).

To this point, Benedetti and Kostoyevsky examine the lifespan of startups that undertake ICOs and find that their survivorship is low, determining that only 44.2% of startups survive after 120 days from the end of their ICOs (2018). Their research suggests that the rush for ICOs was a digital incarnation of the Tulip Mania that overran Europe in the early seventeenth century.

Both quantitative empirical (see Adhami et al. 2018; Fisch 2019; Catalini and Gans 2018; Howell et al. 2018) and qualitative approaches (Chohan 2017a–f, 2018a–c) have been deployed to situate the nature and purpose of ICOs within a traditional financial "language," be it through signalling theory (Fisch 2019), quantity theory of money (Conley 2017), or various other lenses. Yet traditional finance theorizations do not quite fully capture the lived-experience or chaotic spur of the ICO as investment vehicle and subculture. The space may be in fact described as a "wild west," where independent groups have posted alluring suggestions of projects, even without significant detail, and tempted small-scale investors to dip into the supposed prize, only to be left high-and-dry when the ICO's profits fail to materialize. It is in that gap of praxis that concerns over ICO risks have caused alarm in the regulatory and oversight space.

3 The Wild West of ICOs

When the popularity of cryptocurrencies among a wider public began to soar (circa 2016), so too did the amount of ICOs promising ever greater returns to investors. Benedetti and Kostovetsky (2018) describe this phenomenon as a digital reiteration of the Tulip Mania which engulfed Europe in the early decades of the seventeenth century, and Chohan remarks that at the peak of cryptocurrency hype in November, 2017 there were already more than 50 ICOs taking place every month (Chohan 2017d).

Prominent early ICO token sales included Mastercoin in July, 2013, and both Ethereum and Karmacoin in 2014, along with what were termed more "mainstream" ICOs (i.e. more in line with a traditional and institutional investor base) occurring with Kik in September 2017. Even at these early stages, fraudulent practices were being observed, as when Kik faced a phishing scam through a false online link (URL). Yet investor interest remained heavy, as when the web browser Brave's ICO generated $35 million in less than 30 s (Chohan 2017d).

Nevertheless, although a cumulative analysis of ICO volumes showed that capital-raising through ICOs was significant (Satis Group 2017; Adhami et al. 2018; Chohan 2017d), large numbers of ICOs resulted in "substantial scam-artistry, phishing, Ponzi schemes, and other shenanigans" (Chohan 2017d, p. 5). But the scale of such practices is truly frightening. According to one study which examined the lifecycle of ICOs from the initial proposal to the final phase of trading on a crypto-exchange, more than 80% of ICOs emitted in 2017 were scams (Satis Group 2017), amounting in value terms to more than US$1 billion (value estimates of the total capital raised in that year was $11 billion). For 2018, another ICO advisory firm estimated that, for more $20 billion in capital raised from 789 ICOs, the 10 largest ICO scams swindled a combined amount of more than $700 million (Fortune Jack 2018). Various open-access online databases such as Deadcoin began to tabulate a large litany of fake and fraudulent cryptocurrencies, with labels such as "scam," "pyramid scheme," "hack," "disaster," and the pejoratively titled "shitcoin," (Deadcoin 2019). As of this writing, the Deadcoin graveyard enlists hundreds of false, fraudulent, or defunct coins. Benedetti and Kostoyevsky have determined that only 44.2% of startups survive after 120 days from the end of their ICOs (2018). The larger ICO scams by value, as of this writing, include Pincoin and iFan's colossal $660 million dollar swindle, along with Plexicoin, Centra Tech, Bitconnect, Bitlicense, and Bitcard (Fortune Jack 2018; see Bitconnect analysis in Chohan 2018a; Bitlicense analysis in Chohan 2018c).

The ambit of nefarious practices within the under-regulated space of ICOs has been large, with damaging consequences for the public, for the reputation of cryptocurrencies, and even for regulators towards whom fingers were unjustly pointed once real losses were being incurred. For so recent a domain, cryptocurrencies have indeed found substantial presence in the public discourse, in news media, and in the online forums where vibrant discussion has taken place (Chohan 2017a). A wide gamut of attitudes towards cryptocurrencies persists even today, and this is reflected in the regulatory attitudes of various jurisdictions as well (Chohan 2017b).

On one hand, cryptocurrencies are seen as a burst of innovation in a world where even digital technologies have come to stand as monopolistic structures (e.g. Google for search engines, Facebook for social media interactions). Cryptocurrencies were being lauded as a surge of citizen-driven innovation (Chohan 2019d) in the seemingly ossifying digital world of giant corporations.

Yet on the other hand, the lackadaisical levels of due diligence, the wildly inflated promises of transformation, and the quintessential human traits of greed and "fear of missing out" (colloquially termed "FOMO"), all conspired towards severe monetary exploitation of the first order. After all, there are reasons that financial regulation exists. There are indeed reasons why entities that wish to emit securities must comply with a long list of requirements, imposed upon by bodies that are created with a mandate to function in the public interest (Clayton and Giancarlo 2018; Giancarlo 2018). Without such regulatory bodies, the downswings of the free market are much more ruthless, and market failures are much more drastic.

By that line of argument, those investors who dealt with suspicious ICOs of their own accord must be answerable for their own choices. Under no compulsion did these individuals invest their own money into the seemingly endless rhetorical promises of the cryptocurrency realm. This is why Benedetti's and Kostovetsky's (2018) comparison to the Tulip Mania of the European Renaissance is perhaps apt, for as with the short-lived boom in the price of a whimsical commodity (flowers), coupled with the insatiable appetites of investors, great fortunes were lost in but an instant.

After all, there was far less complaint about ICOs when cryptocurrency prices were at their zenith (see Chohan 2018a–c). Rather, it was when the prices declined that the furore of losing investors spread across the online forums and into the public sphere. For all those proponents of cryptoanarchist attitudes towards the freedom to invest, many in fact would demand recourse from the existing structures of financial accountability. Seldom in the good times, but often in the bad times, would the weaknesses of the cryptocurrency space as unregulated domain be articulated thus.

Indeed, the scope of widespread financial abuse through ICOs came to jeopardize the reputation of the space as a whole (Chohan 2019a–c), with many small- and large-scale investors demanding recourse and recovery of funds. Given that the inherent design of cryptocurrencies is to situate them outside the traditional financial architecture (Fisch 2019; Howell et al. 2018), such demands pose a dilemma for regulatory authorities around the world (GAO 2014), initially due to sheer bewilderment at the meteoric rise of the sector (see Chohan 2017b, 2019c), but since then due to the need to strike a balance between fostering innovation and imposing accountability (see Chohan 2019d, e). Those issues are discussed in the following section.

4 Regulatory Responses to ICOs

ICOs have "low contributor protection, a limited set of available information, [almost] no supervision by public authorities, and [almost] no relevant track record for proponents," (Adhami et al. 2018, p. 73). It was remarked early on that the divergences in international regulatory responses to cryptocurrencies were quite stark (Chohan 2017b; Adhami et al. 2018, p. 65–66), ranging from outright banishment of cryptocurrencies from the financial architecture of some countries, to the enthusiastic embrace of others. However, the picture grew more nuanced since 2018, when a general price decline in crypto-instruments led to a more vocal chorus of disenchantment with the promise of cryptocurrencies (Chohan 2019a–c). As noted in earlier sections, even in cases of legitimate ICOs, funded projects are typically in a high-risk early stage of development, with considerable downside potential for investors (Conley 2017; Howell et al. 2018; Li and Mann 2018).

With that in mind, Chohan has argued that OECD countries, with the Securities and Exchange Commission (SEC) and Commodities and Futures Trading Commission (CFTC) of the United States taking the lead, have attempted to strike a balance between the principles of "innovation" and "accountability" (Chohan 2019e). The US Securities and Exchange Commission (SEC) has issued explicit warnings to investors to be highly cautious against scammers using ICOs, particularly in the colloquially termed "pump and dump" schemes, where capital is fleetingly raised and then immediately dumped in exchange for other (more established) instruments at a profit, all within a very brief interval (Clayton 2017). The UK Financial Conduct Authority has also warned that ICOs are very high risk and speculative investments, are scams in some cases, and often offer no protections for investors (Chohan 2017d). Australia's regulator (ASIC) has issued guidance (September, 2017) stating that the legality of an ICO is dependent on the specific circumstances, on a case-by-case basis.

An even more reticent attitude has been expressed by financial regulators in China, where seven regulatory agencies officially banned all ICOs within the People's Republic, and they demanded that the proceeds from all past ICOs be refunded to investors or face being severely punished according to the law (Li and Mann 2018; Lee et al. 2018). This decision is being reconsidered, as of this writing. The Chinese context is important because ICOs had raised nearly $400 million from about 100,000 investors prior to the ban. However, more recent statements from Chinese regulators have stated that the ICO ban is intermittent, pending a more systematic regulatory framework.

A similar situation, and a more surprising one, has emerged as of this writing in Switzerland. Although Switzerland was previously viewed as a jurisdiction amenable and friendly to ICOs, in September, 2017 the Swiss Financial Market Supervisory Authority announced an investigation of an unspecified number of coin offerings to examine whether they complied with Swiss regulations (Chohan 2017d). A strong line has also been taken by regulators in South Korea, where the Financial Services Commission prohibited ICOs in September 2017 and promised "stern penalties" for violations (Li and Mann 2018; Lee et al. 2018).

Given the recency of the ICO phenomenon, many important jurisdictions have yet to issue regulatory guidelines, of this writing. However, more comprehensive guidance has been issued by Hong Kong, New Zealand, Australia, Gibraltar, and the UAE. In Hong Kong, the Securities and Futures Commission released a statement (September 2017) explaining that tokens may constitute securities for purposes of the legal framework (Securities and Futures Ordinance), in which case dealing in such tokens would be a regulated activity under Hong Kong law. In New Zealand, the Financial Markets Authority (FMA) released guidelines on the current regulatory environment in regards to ICOs in October, 2017. In Gibraltar, the government published regulation establishing a framework for regulated DLT (Distributed Ledger Technology) companies, which would encompass ICOs and subject them to financial controls and standards; which entered into effect on January 1, 2018. In the UAE, the Abu Dhabi Global Market issued official guidance on ICOs in October, 2017.

Nevertheless, the richest experience with the regulation of cryptocurrencies and their ICOs has come from the SEC and CFTC in the United States (Chohan 2019d). As far back as 2017, the Chairman of the SEC had insisted upon investors the need to exercise caution given: the financial dangers of being misled by fraudulent cryptocurrency agents; the international nature of cryptocurrency fund flows; and the emphasis on the substance of transactions rather than their form (Clayton 2017).

In his remarks, Chairman Clayton noted that advocates were claiming that tokens issued on cryptocurrencies were more of a "utility" than a security, and responded that this emphasized the form of tokens, rather than their substance (2017). Instead, a nuance was put forth in that "these [ICOs] can take many different forms, and the rights and interests a coin is purported to provide the holder can vary widely," (Clayton 2017). "By and large," he observed, the structures of ICOs "involve the offer and sale of securities and directly implicate the securities registration requirements and other investor protection provisions," and that these laws "provide that investors deserve to know what they are investing in and the relevant risks involved," (Clayton 2017).

Even at the rudimentary level of classifying cryptocurrencies, these institutions have deliberated greatly and arrived at rulings that have shaped the international ICO space. This is not a trifling matter, and has been as much of a philosophical problem as a technical one. If cryptocurrencies were treated as property, they would be regulated in the US by the Internal Revenue Service (IRS). If they were securities, they would fall under the Securities and Exchange Commission (SEC). If they were commodities, they would come under the Commodities Futures Trading Commission (CFTC). The determination of their asset class status would therefore have important ramifications for ICOs.

The chairman remarked that the SEC's Division of Enforcement would "police this area vigorously and recommend enforcement actions against those that conduct initial coin offerings in violation of the federal securities laws," (Clayton 2017). A substantial series of enforcement actions have since been taken against ICO issuers who have not complied with securities regulation. However, the SEC has also presented a more nuanced treatment of cryptocurrencies and ICOs by separating the purview of the space into both securities and commodities (see Clayton and Giancarlo 2018).

In June, 2018, a joint statement was issued by the chairmen of the SEC and CFTC (see Clayton and Giancarlo 2018). They noted that "many have identified [cryptocurrencies and their ICOs] as the next great driver of economic efficiency. Some have even compared it to productivity-driving innovations such as the steam engine and personal computer," (2018). At the same time, the chairmen emphasized closer cooperation between their agencies (Clayton and Giancarlo 2018), and a closeness of views in approaching the cryptocurrency space. Furthermore, they spoke to the need for regulations to strike a balance when they "set and enforce rules that foster innovation, while promoting market integrity and confidence," (Clayton and Giancarlo 2018). In a later interview, CFTC Chairman Giancarlo insisted that the purpose of regulation was not to stifle ICOs, but to protect investors, stating the following: "I think that cryptocurrencies are here to stay. I think that there is a future for them. [But] I am not sure if they will ever come to rival the dollar or other hard currencies, but there is a whole section of the world that is hungry for functioning currencies, [like Bitcoin]," (Giancarlo 2018).

The SEC and CFTC are thus leading the pack of international regulators in protecting investors and regularizing ICOs as an investment vehicle. Their approach is likely to influence regulators around the world, and so even a disjointed international regulatory landscape is likely, through isomorphic pressures, to come to the standards set by the US SEC and CFTC. Whether this isomorphism will be mimetic or normative cannot be said at this early juncture. Yet a growing public pressure in the wake of volatile (and declining) prices of cryptocurrencies, followed by a massive scale of fraudulent activity, is likely to pressurize regulators around the world to respond (see also GAO 2014). After all, it has been suggested that increased regulation of ICOs should encourage institutional investors to invest along more stable horizons, and in larger volumes, over more instruments (Chohan 2017a, d). With strong accountability, the ICO market can thrive, and the SEC notes that ICOs can provide fair and lawful investment opportunities (Clayton 2017).

5 Conclusion

ICOs have mostly occurred in the online realm that lies beyond regularized and traditional finance, devoid of the structures of financial regulation which allow for capitalism to function in a more stable and lawful manner (Chohan 2017b, 2019a, b). Instead, ICOs lie philosophically within cryptoanarchist thought, which seeks to cultivate decentralized, autonomous, and voluntary exchange among individuals in a manner that protects their identities and therefore their risk of persecution by structures of government (Chohan 2017f). This creates a conundrum for those adherents of cryptocurrencies who wish for these instruments to remain "free" from traditional oversight. For cryptoanarchism, as with anarchism itself (see Wolff 1998; Marshall 2009), there are utopian expectations of human beings that remain wanting (at least thus far in the human experience), including a selflessness and trust between groups of people who will demonstrate respect and consideration in an effort to come to mutual aid. In an anonymous world of trading bits of code as monetized instrument, even as it may be nominally "trustless," issues of trust have indeed surfaced, and often bitterly so.

Indeed, the massive frauds of the ICO space over the last few years have compelled roiled investors run to traditional financial regulators for recourse, thereby challenging the praxis of utopian cryptoanarchist ideals. It has been observed that, while the prices of cryptocurrencies were rising circa late 2017, there was a much more vocal celebration of the decentralized nature of cryptocurrencies, removed from the control and regulation of the traditional banking system (Chohan 2017a, b). However, as the prices plummeted and heavy losses were incurred among members of the public in early 2018, there was a great deal more hue-and-cry about the low levels of accountability in the cryptocurrency space (Chohan 2017d).

Whereas the hard proponents of ICOs argue that "innovation" is what is at stake here, a lack of sufficient ICO accountability and oversight have raised a public furore which is now being met, however haphazardly, by traditional regulatory authorities. Indeed, a very vocal argument is being made that ICOs do require accountability and regulation in the traditional sense. In any case, the recency of the ICO phenomenon necessitates both academic and practitioner considerations of the risks, regulation and accountability mechanisms that are self-reinforcing and dynamic, in the same way that the innovation of ICOs is.

The literature on ICOs therefore requires much more development so as to confront the rapid changes occurring in the practitioner sphere. This concluding section enumerates some of those areas of future inquiry. First, a comparison of the gains and risks of raising capital through cryptocurrency mechanisms rather than fiat money, in both regulated and unregulated jurisdictions, warrants greater (particularly empirical) attention (see example in Catalini and Gans 2018). Second, the scope of regulation, as securities or as commodities, still poses a challenge to regulators around the world, although significant progress has been made in the United States (Chohan 2019d). Related to this point, we might ask whether investor protections against fraud must be increased in both the primary and secondary markets. If so, would cryptocurrencies have any "crypto" element truly left?

The question of how the balance between "accountability" and "innovation" should be struck differs between countries' regulatory systems and commercial culture. So a third area of suggested research would be whether an international, unified system of financial oversight could or should include cryptocurrencies and their ICOs. A fourth area of enquiry would be on the demand side: why are so many online contributors still eager to transfer large sums of traditional many to fund ICOs? Behavioural economics may have much to contribute in that regard. A fifth area would speak to the "value" created for the public through ICOs, and how public managers (regulatory bodies) can participate in that value creation process when it is in fact being driven by civil society and individuals (see initial work in Chohan 2019e).

In sum, there are powerful emotive elements that have surfaced alongside the meteoric rise of cryptocurrencies as both asset class and cultural phenomenon. The wild west of ICOs has been a particularly troublesome issue in cryptocurrencies, ultimately de-legitimizing the entire space to at least some degree. The vociferous demands for accountability and recourse have force regulators to step in to the space, particularly as the losses grow, but also as risks of further fraud, money laundering, and theft increase. Their response was initially slow and reactive, but some bodies such as the SEC and CFTC are taking important and measured steps towards improving the scope of regulation in the field.

This chapter has sought to emphasize the need for regulation and oversight of the ICO domain. The downside risk of losses is immense given the propensities for nefarious activities that exist whenever money flows in the shadows. Although this does challenge the ideals of cryptoanarchism which adherents of cryptocurrencies invoke, there is a strong case for better and tighter regulation of these instruments. The purpose of that regulatory effort must be to strike a balance between "innovation" on one hand and "accountability" on the other. By such an approach, regulation can help to bolster the credibility of ICOs as a vehicle for raising capital quickly to fund technical digital projects, while also mitigating some of the weaknesses of ICOs, so as to better leverage its strengths towards value creation and innovation.

References

Adhami S, Giudici G, Martinazzi S (2018) Why do businesses go crypto? An empirical analysis of initial coin offerings. J Econ Bus 100:64–75

Benedetti H, Kostovetsky L (2018) Digital tulips? Returns to investors in initial coin offerings. Returns to investors in initial coin offerings, May 20, 2018

Catalini C, Gans JS (2018) Initial coin offerings and the value of crypto tokens (No. w24418). National Bureau of Economic Research

Chohan UW (2017a) Cryptocurrencies: a brief thematic review. Notes on the 21st century. Critical Blockchain Research Initiative (CBRI) Working Papers. https://papers.ssrn.com/sol3/papers.cfm?abstract_id=3024330

Chohan UW (2017b) Assessing the differences in Bitcoin & other cryptocurrency legality across national jurisdictions. Notes on the 21st century. Critical Blockchain Research Initiative (CBRI) Working Papers. https://papers.ssrn.com/sol3/papers.cfm?abstract_id=3042248

Chohan UW (2017c) A history of Bitcoin. Notes on the 21st century. Critical Blockchain Research Initiative (CBRI) Working Papers. https://papers.ssrn.com/sol3/papers.cfm?abstract_id=3047875

Chohan UW (2017d) Initial coin offerings (ICOs): risks, regulation, and accountability. Notes on the 21st century. Critical Blockchain Research Initiative (CBRI) Working Papers

Chohan UW (2017e) The double spending problem and cryptocurrencies. Notes on the 21st century. Critical Blockchain Research Initiative (CBRI) Working Papers. https://papers.ssrn.com/sol3/papers.cfm?abstract_id=3090174

Chohan UW (2017f) Cryptoanarchism and cryptocurrencies. Notes on the 21st century. Critical Blockchain Research Initiative (CBRI) Working Papers. https://www.ssrn.com/abstract=3079241

Chohan UW (2018a) Bitconnect and cryptocurrency accountability. Notes on the 21st century. Critical Blockchain Research Initiative (CBRI) Working Papers. https://papers.ssrn.com/sol3/papers.cfm?abstract_id=3131512

Chohan UW (2018b) The problems of cryptocurrency thefts and exchange shutdowns. Notes on the 21st century. Critical Blockchain Research Initiative (CBRI) Working Papers. https://papers.ssrn.com/sol3/papers.cfm?abstract_id=3131702

Chohan UW (2018c) Oversight and regulation of cryptocurrencies: BitLicense. Notes on the 21st century. Critical Blockchain Research Initiative (CBRI) Working Papers. https://papers.ssrn.com/sol3/papers.cfm?abstract_id=3133342

Chohan UW (2019a) The key man problem in cryptocurrencies? Case of QuadrigaCX. Notes on the 21st century. Critical Blockchain Research Initiative (CBRI) Working Papers. https://papers.ssrn.com/sol3/papers.cfm?abstract_id=3329573

Chohan UW (2019b) Cryptocurrencies and financial conduct. Notes on the 21st century. Critical Blockchain Research Initiative (CBRI) Working Papers. https://papers.ssrn.com/sol3/papers.cfm?abstract_id=3329929

Chohan UW (2019c) Are cryptocurrencies truly 'trustless'? Notes on the 21st century. Critical Blockchain Research Initiative (CBRI) Working Papers. https://papers.ssrn.com/sol3/papers.cfm?abstract_id=3331544

Chohan UW (2019d) Public value and citizen-driven digital innovation. Information Polity

Chohan UW (2019e) Public value without public managers? Decentralized autonomous organizations and public administration. Information Polity

Clayton J (2017) Statement on cryptocurrencies and initial coin offerings. In: Securities and Exchange Commission (SEC) Statements. SEC, Washington, DC

Clayton J, Giancarlo JC (2018) Joint statement on cryptocurrencies and initial coin offerings. Securities and Exchange Commission (SEC) & Commodity Futures Trading Commission (CFTC). Washington, DC

Conley JP (2017) Blockchain and the economics of crypto-tokens and initial coin offerings (No. 17-00008). Vanderbilt University Department of Economics

Deadcoin (2019) Database on dead coins. https://deadcoins.com/

Fisch C (2019) Initial coin offerings (ICOs) to finance new ventures. J Bus Ventur 34(1):1–22

Fortune Jack (2018) Study on ICO scams in 2018. https://fortunejack.com/

Giancarlo JC (2018) Interview with fast money on cryptocurrency regulation. Securities Industry and Financial Markets Association (SIFMA). Washington, DC

Government Accountability Office [GAO] (2014) Virtual currencies: emerging regulatory, law enforcement, and consumer protection challenges. GAO-14-496. Washington, DC

Howell ST, Niessner M, Yermack D (2018) Initial coin offerings: financing growth with cryptocurrency token sales (No. w24774). National Bureau of Economic Research

Kaal WA, Dell'Erba M (2017) Initial coin offerings: emerging practices, risk factors, and red flags. Verlag CH Beck (2018), 17–18

Lee J, Li T, Shin D (2018) The wisdom of crowds and information cascades in FinTech: evidence from initial coin offerings

Li J, Mann W (2018) Initial coin offering and platform building

Marshall P (2009) Demanding the impossible: a history of anarchism. PM Press, Oakland

Satis Group (2017) Study on ICO scams in 2017. http://satisgroup.io

Venegas P (2017) Initial coin offering (ICO) risk, value and cost in blockchain trustless crypto markets

Wolff RP (1998) In defense of anarchism. University of California Press, Berkeley

Cryptocurrencies and Risk Mitigation

Haifa Amairi, Boushra El Haj Hassan, and Ahlem Zantour

Abstract Since their surge in the last decade cryptocurrencies have gained considerable attention in financial markets, and in academic research. Scholars and practitioners are showing interest in the role of cryptocurrencies as part of investors' risk management strategies. Understanding how the returns of different cryptocurrencies, and the associated volatilities, relate to the returns and volatilities of other assets (including other cryptocurrencies, stocks, commodities, and bonds, among others) is crucial to derive conclusions regarding the potential hedging and diversification advantages they could offer to investors' portfolios. The notion of volatility transmission, its intensity and direction, is of importance in explaining the risk management benefits that could stem from adding a specific asset, such as cryptocurrencies, to an existing portfolio.

1 Introduction

When considering the risk-related effect stemming from adding a security to a portfolio, there are many possible classifications for the role such security could play. For instance, a security can be categorized as a hedging asset, a diversifier, or a safe-haven, depending on its properties. Bauer and Lucey (2010) provide detailed definitions to distinguish between these different types. For instance, a diversifier is an asset that is on average positively, but not perfectly, correlated with another asset

H. Amairi
École Supérieure de Commerce, Tunis, Tunisia

Telfer School of Management, University of Ottawa, Ottawa, ON, Canada
e-mail: hamai100@uottawa.ca

B. El Haj Hassan (✉)
Telfer School of Management, University of Ottawa, Ottawa, ON, Canada
e-mail: belha068@uottawa.ca

A. Zantour
Telfer School of Management, University of Ottawa, Ottawa, ON, Canada

IHEC Carthage, Tunis, Tunisia

S. Goutte et al. (eds.), *Cryptofinance and Mechanisms of Exchange*, Contributions to Management Science, https://doi.org/10.1007/978-3-030-30738-7_11

or portfolio. A hedge is an asset that is, on average, uncorrelated or negatively correlated with another asset or portfolio. As these definitions explicitly specify, the correlation properties are only required to hold on average for an asset to be classified as a hedge or as a diversifier. In other words, a hedge or a diversifier might not enable loss reduction under extreme market conditions or turmoil. An asset with the properties of a hedge or a diversifier under regular market conditions could exhibit completely different (correlation) properties under extreme adverse market conditions. On the other hand, a safe-haven, by definition, is an asset that is uncorrelated or negatively correlated with another asset or portfolio in times of market stress; for instance, under extreme adverse market conditions (i.e. bearish eras), the price of such an asset increases when the price of the other assets or portfolio decreases. In quiet (regular) periods, such an asset might behave differently, and possibly exhibit positive correlation with the same asset or portfolio. Baur and McDermott (2010) highlight the distinguishing feature of a hedge compared to a safe-haven, which is the length of the effect. While the correlation property of a hedge should hold on average, the key property of the safe-haven is required to hold during certain periods, such as financial crises. The correlation properties of the same asset with relation to another asset or portfolio could be different during a crisis, than during periods exhibiting regular market conditions. Another aspect that is crucial to be considered by investors is the distinction between a strong hedge, that is negatively correlated with another asset or portfolio, and a weak hedge, that is, on average, uncorrelated with the other asset(s) or portfolio. Whereas a strong hedge might enable an investor to enjoy significant positive returns when the other asset(s) or portfolio suffers from negative returns, the same might not hold for a weak hedge.

2 Cryptocurrency Market Efficiency

An important notion relating to the return and risk relation is that of market efficiency. The efficient market hypothesis, which has been developed by Fama (1970), has been the basis for many foundational theories in finance. A market is said to be efficient if security market prices reflect all available information. Given that security prices are assumed to incorporate all relevant information, no one can persistently beat the market. There are three forms of market efficiency; (1) the weak form, (2) the semi-strong form, and (3) the strong form which respectively refer to the inability of an investor to outperform the market and generate excess abnormal returns (given the level of risk undertaken) based on (1) information on previous security prices, (2) any public information available, and (3) any public or private information available.

Many recent studies have focused on exploring the efficiency of cryptocurrency markets. By examining the liquidity of 456 different cryptocurrencies, Wei (2018) found that higher liquidity is associated with higher efficiency, referring to a lower predictability of future returns. Based on Wei's findings, the Bitcoin market shows signs of efficiency, whereas the returns of other cryptocurrencies exhibit signs of auto-correlation and non-independence and thus their markets are not efficient.

Brauneis and Mestel (2018) also found liquid cryptocurrencies' markets to be more efficient. On the other hand, Charfeddine and Maouchi's (2018) examined the Long-Range Dependence (LRD) behavior of the returns and volatilities of four cryptocurrencies, namely, Bitcoin, Litecoin, Ripple, and Ethereum. Their findings confirmed the inefficiency of all the cryptocurrencies covered by their study, except for Ethereum. Caporale et al. (2018) also concluded based on their study that the Bitcoin, Ripple, Litecoin, and Dash markets are inefficient. Nadarajah and Chu (2017) and Urquhart (2016) added to the evidence on the inefficiency of the Bitcoin market. However, when splitting their sample into subsample periods, Urquhart (2016) found that Bitcoin became more efficient in later periods, and thus concluded that this currency might be in the process of moving towards efficiency. Vidal-Tomás and Ibañez (2018) examined the semi-strong efficiency of Bitcoin in the Bitstamp and Mt.Gox markets to explore how this currency is affected by its own events and the monetary policy. In line with Urquhart's (2016) conclusions, they found that the Bitcoin market is becoming more efficient over time, and that it is not affected by monetary policy news.

Beneki et al. (2019) used innovative VAR methodologies to investigate the volatility transmission from Ethereum to Bitcoin (first and second cryptocurrencies in terms of market capitalization and trading volume) throughout time. They noted a delayed response of Bitcoin's volatility in response to volatility shocks to Ethereum's returns, which is interpreted as signs of inefficiency in the Bitcoin markets. Given that such public information is shown to take time to be incorporated into Bitcoin prices, a profit-making opportunity could exist for investors to benefit from.

3 Risk Mitigation Using Cryptocurrencies

Given that Bitcoins are the first cryptocurrencies to emerge in the market and the one with the largest market capitalization, these have been the most explored by existing studies compared to other cryptocurrencies. A significant thread of the literature on cryptocurrencies studies the role of Bitcoins as a safe-haven, a diversifier, or a hedging asset vis-à-vis other assets, as well as other cryptocurrencies (i.e. Bouri et al. 2017a–c, 2018; Dyhrberg 2016a, b; Brière et al. 2015; Feng et al. 2018; Urquhart and Zhang 2018; Fang et al. 2019). A crucial factor supporting the hedging ability of Bitcoin relates to its independence with relation to economic and financial developments (Polasik et al. 2015; Bouri et al. 2017c; Guesmi et al. 2018), and negative or weak positive (or lack of) correlation with conventional assets (Bouri et al. 2017c; Baur et al. 2015, 2018; Yermack 2013; Dyhrberg 2016a, b; Corbet et al. 2018; Ji et al. 2019; Guesmi et al. 2018; Sun et al. 2018).

On the other hand, there are studies arguing that the little intrinsic value (Yi et al. 2018) and the high volatility in Bitcoin prices (Molnár et al. 2015; Wong et al. 2018), due to the speculative nature of this market, could weaken the role of this cryptocurrency as part of a diversification strategy (i.e. Cheah and Fry 2015). The decentralization and the fixed supplies of cryptocurrencies, including Bitcoin, makes them more susceptible to short-term price fluctuations (Berentsen and Schar 2018). Using a GARCH (1.1) model, Corbet et al. (2017) examined the effects of international monetary policy changes on the volatility of bitcoin returns, and found these effects to be significant, thus questioning the widespread claims of Bitcoin's independence vis-à-vis government policies. This result entails resemblance between Bitcoin and other store of value assets and currencies, and have implications on the consideration of Bitcoin for hedging and diversification purposes.

According to Guesmi et al. (2018), Bitcoin possesses hedging abilities in various financial markets, and using it as part of a diversification strategy could contribute to reducing the risk of the investment. Such a conclusion is supported by the evidence of uncorrelation of Bitcoin with traditional asset classes provided in Baur et al. (2018); this evidence stemmed from a correlation analysis involving Bitcoin returns and the returns of traditional asset classes. Works by Dyhrberg (2016a, b) also provided evidence on the hedging capabilities of Bitcoin; using the standard GARCH and exponential GARCH (EGARCH) models, they find Bitcoin to have hedging properties and advantages as a medium of exchange, because of the similarities it shares with the gold and dollar markets (Dyhrberg 2016a). Then, in another study, they provided evidence on Bitcoin's role as a hedge against UK equities and the US dollar (Dyhrberg 2016b). There are other studies that have argued for the benefits of including Bitcoin into diversified portfolios by improving the risk-return trade-off (i.e. Halaburda and Gandal 2014; Eisl et al. 2015; Chen and Vivek 2014; Brière et al. 2015).

Evidence on the dynamic hedging abilities of Bitcoin against many stock indices was provided in a study by Chan et al. (2019); these indices included Euro STOXX, Nikkei, Shanghai A-Sharpe, S&P 500, and the TSX index. In this study, they have used GARCH models and constant conditional correlation models for daily, weekly, and monthly returns covering the period from October 2010 until October 2017. The movements of the daily returns have been decomposed into high, medium, and low frequency movements using the frequency dependent model. The insignificant correlation exhibited between Bitcoin and the indices' returns over the daily and weekly horizons undermines its hedging abilities against market risk in the short run. However, for monthly horizons, the hedging abilities of Bitcoin improve significantly, due to the significantly negative correlations exhibited towards the indices considered.

Using a dynamic conditional correlation GARCH (DCC-GARCH) method, Bouri et al. (2017c) found support for the role of Bitcoin as a diversifier rather than a hedge against stock indices, bonds, oil and gold. Their evidence also shows Bitcoin to possess strong hedging abilities against the commodity index and that it can act as a strong safe-haven against extreme down movements in Chinese stocks and Asia Pacific stocks. However, Bitcoin's hedging and safe-haven properties differed across time. For instance, the hedging abilities of Bitcoin against the commodity index exhibited through daily data vanished with weekly data, and the hedging properties against the Japanese stocks for daily data faded with weekly data. The hedging and safe-haven properties against the Chinese stocks revealed through weekly data were not present with daily data. In addition, for the Asia Pacific stocks,

Bitcoin played a hedging role based on daily data, and progressed to act as a safe-haven based on weekly data.

In another study, Bouri et al. (2017a) used the asymmetric GARCH method to identify how Bitcoin's risk-reduction abilities have changed after its price crash in 2013. They found that, while before the price crash Bitcoin had similar safe-haven properties as gold, these properties have vanished in the subsequent periods. In addition, they reported evidence that adding Bitcoin to US equity portfolios enables risk reduction.

The findings of Dyhrberg (2016a, b) also support the role of Bitcoin in reducing risk (like gold) through its hedging capabilities, specifically against the Financial Times Stock Exchange (FTSE) Index and the US dollar. Therefore, as part of her conclusions, she classified Bitcoins as a hybrid between a currency and a commodity. By focusing on emerging stock markets, Guesmi et al. (2018) were able to highlight the role of Bitcoin as a diversifier, even in portfolios including oil and gold. Selmi et al. (2018) argue for similarities between Bitcoin and gold in mitigating portfolio risk related to fluctuation in oil price movements.

By focusing on the three cryptocurrencies that account together for more than 40% of the total cryptocurrency market capitalization (Bitcoin, Litecoin, and Ripple), Wong et al. (2018) found that Bitcoin and Litecoin have negative or zero correlations with other asset classes, and can thus act as hedges, and that Ripple exhibits diversification properties (slight positive correlations with the other asset classes). In contrast with Bouri et al. (2017c), Wong et al. (2018) found that Bitcoin can act as a hedge against stocks (S&P500) due to the presence of a significant negative correlation between the two, whereas the former study found Bitcoin to be a diversifier in relation to the S&P500. A possible explanation of these differing results, according to Wong et al., is due to the difference in the sample periods covered by the two studies. Corbet et al. (2018) conducted a spillover analysis in which they examined the relation between Bitcoin, Ripple, and Litecoin, and other asset classes. Their findings revealed that these cryptocurrencies are rather immune to external market shocks, therefore they can be considered to be useful as diversifiers and safe-havens over short time horizons, and that when added to a portfolio, they result in an enhanced risk-return trade-off.

4 Hedging Cryptocurrency Investments

In addition to using cryptocurrencies to reduce risks associated with investing in securities (such as stocks, bonds, among others) and commodities, cryptocurrencies can also be used to mitigate risks resulting from investing in other cryptocurrencies. The heterogeneity exhibited across the risk levels of different cryptocurrencies (Gkillas and Katsiampa 2018; Brauneis and Mestel 2018) could prove to be useful when it comes to the diversification benefits they could entail (Antonakakis et al. 2019). Exploring the risk mitigation effects stemming from combining cryptocurrencies in a portfolio requires examining return and volatility connectedness or spillovers among these cryptocurrencies to get informed on the information transmission mechanism involved (Yi et al. 2018). For instance, a weak connectedness across cryptocurrencies could present diversification and hedging opportunities for investors (Ji et al. 2019), whereas a higher level of connectedness and spillovers among cryptocurrencies would be expected to limit the hedging and diversification benefits resulting from combining them into a portfolio (Yi et al. 2018).

Beneki et al. (2019) investigated the volatility transmission and hedging properties between Bitcoin and Ethereum throughout time to explore the existence of trading strategies that could result in abnormal profits given the risk levels undertaken by investors. Using innovative VAR methodologies allowed them to examine responses of time-varying volatilities of Bitcoin to time-varying volatilities of Ethereum. They documented a delay in the response of Bitcoin to volatility shocks to Ethereum returns, which was interpreted as inefficiency in the Bitcoin market. This delay could present opportunities for speculation and profit-making for investors. In addition, they found that, during the first half of 2017, when prices of both Bitcoin and Ethereum increased, these two cryptocurrencies had a near-zero correlation, so they acted as diversifiers rather than hedges. The hedging abilities of these cryptocurrencies decreased significantly in later periods and during periods of increased policy uncertainty (Beneki et al. 2019). Findings in Corbet et al. (2018) also outline the diversification benefits that emerge from investing in Bitcoin, Ripple, and Litecoin, specifically for short-term oriented investors.

In a study by Borri (2019) that considered Bitcoin, Ether, Ripple, and Litecoin, he used a CoV aR methodology to examine the conditional tail-risk in the markets of these cryptocurrencies. He found that, despite the high correlations in the returns of these cryptocurrencies, investing in a portfolio of cryptocurrencies allows the reduction of idiosyncratic risk and offers a better risk-adjusted performance and conditional returns than investing in individual cryptocurrencies. Ether, Litecoin and Ripple seemed to be vulnerable to the tail-risk of Bitcoin, whereas Bitcoin seemed to be more resilient to shocks to the returns of the other cryptocurrencies considered. By examining the co-movement between dollar returns on these cryptocurrencies and other global assets, such as gold and US equity, both unconditionally and conditional on these assets being in a state of distress, he found that cryptocurrencies are poorly correlated with, and not exposed to tail-risk with respect to global assets. He concluded that cryptocurrency portfolios could represent hedging properties to investors and offer attractive returns. In a similar vein, Antonakakis et al. (2019) used a TVP-FAVAR connectedness approach to examine the transmission mechanism in the cryptocurrency markets. They have explored co-movements in the markets of the top nine currencies (by virtue of their market capitalization), and one market factor (that entail 45 additional digital currencies) to capture the main return co-movements in the crypto-market. Their results reveal large dynamic variability (ranging between 25 and 75%) across several cryptocurrencies, and stronger (lower) connectedness during periods of higher (lower) market uncertainty. Their conclusion supports their proposition that higher interconnectedness in the crypto-market facilitates portfolio and risk management techniques. By constructing bivariate dynamic portfolios, their findings suggest that including Bitcoin and Ethereum in a portfolio results in more effective diversification.

Both static and dynamic volatility connectedness among cryptocurrencies have also been investigated in a study by Yi et al. (2018). They have studied eight cryptocurrencies that were selected based on their market capitalization and long trading history (Bitcoin, Ripple, Litecoin, Peercoin, Namecoin, Feathercoin, Novacoin and Terracoin). They also considered key events that may affect their connectedness. They found that connectedness varies cyclically, and that it has exhibited an upward trend since the end of 2016, which prompted them to dig deeper into the period from December 2016 until April 2018. Then, they based their analyses on a network view using the LASSO-VAR approach to explore the volatility connectedness using an expanded sample of 52 cryptocurrencies. The results, both based on the eight cryptocurrencies sample and the 52 cryptocurrencies sample, showed that, despite its significant high market capitalization, Bitcoin is not the dominant player of volatility connectedness in the crypto-market. They also found that the 52 cryptocurrencies are tightly interconnected.

Interestingly, Ji et al. (2019) have reached differing results with regards to the leading role of Bitcoin in terms of volatility spillovers. They have measured connectedness by following Diebold and Yilmaz (2014); they built positive/negative connectedness networks, then they used regression to identify the drivers of the degree of connectedness among the cryptocurrencies studied. By studying the return and volatility spillovers across six large cryptocurrencies during the period from August 2015 until February 2018, they found that Bitcoin is the most influential in terms of volatility spillovers, and that Bitcoin and Litecoin are the leading cryptocurrencies in terms of the effects of shocks to their returns on other cryptocurrencies. Ethereum (the second in terms of market capitalization) is shown to be rather a recipient of spillovers. Connectedness via negative returns seemed to be significantly stronger than via positive ones. They also found that Dash and Ethereum showed very low connectedness, which could justify their use for hedging in the crypto-market. Therefore, cryptocurrencies' market capitalization did not prove to be a primary determinant of the significance of the connectedness effects of a cryptocurrency on other cryptocurrencies. In addition, they documented a positive effect of global financial distress periods on both returns and volatility connectedness in the cryptocurrency market. They attributed this result to the speculative nature and lack of transparency in the crypto-market, which makes it highly volatile; such conditions, along with periods of financial distress would be expected to encourage herding behaviors (Baur et al. 2018; Demirer and Kutan 2006), thus positively affecting connectedness among cryptocurrencies.

Understanding price dynamics in the cryptocurrency market and the interconnectedness among cryptocurrencies is crucial to determine how portfolios' risk can be better managed. An important characteristic that could limit the diversification potential is the existence of systematic structural breaks, which indicates market integration. Canh et al. (2019) analyzed structural breaks and volatility spillovers in seven cryptocurrencies; Bitcoin, Litecoin, Ripple, Stellar, Monero, Dash, and Bytecoin. They have used various econometric models in their study, such as cumulative sum test for parameter stability, Granger Causality test, LM test for ARCH and Dynamic Conditional Correlation MGARCH model. Their findings confirm the existence of structural breaks, and volatility spillovers with strong positive correlations among these cryptocurrencies; correlations between six out of the seven cryptocurrencies considered exceeded 0.4, with the largest correlation existing between Bitcoin and Litecoin with a value of 0.746. They also found the structural breaks to spread from smaller cryptocurrencies to the larger ones; the prices of cryptocurrencies with lower market capitalizations change first, and those of larger cryptocurrencies follow. Evidence points to the significance of the non-diversifiable risk within the cryptocurrency market, which can be due to the existence of common economic factors affecting these cryptocurrencies within a short period of time. The interdependence across cryptocurrencies' prices has been outlined in many other studies, such as Ciaian et al. (2018) and Boako et al. (2019).

Given the high volatility in the prices of cryptocurrencies, the existent evidence on the interconnectedness and correlation among them, and the significant losses that could result should the prices in the cryptocurrency market fall (i.e. price crash in the early 2018), considering the use of other financial assets to hedge against cryptocurrencies' down-side price movement is of value. Pal and Mitra (2019) examined the possibility of Hedging bitcoin using other financial assets; they have considered the S&P500 composite index (to represent stocks), wheat (to represent commodities), and gold (to represent precious metals). Their results revealed that each of these assets can be used as a hedge against cryptocurrencies, with the gold being the strongest hedge, compared to wheat and the S&P500 index, with a hedge ratio of 0.7005 obtained through the Generalized Orthogonal GARCH (GO-GARCH) model. The interpretation of this ratio is that a US$1 long position in Bitcoin can be hedged by a short position in the gold market for 70 cents.

As with other financial assets, hedging cryptocurrencies' risk exposure, can be performed using derivatives, such as futures, forwards, swaps, and options. Research on cryptocurrency derivatives is relatively scarce, as this market is still in its early stage of development. For instance, the first trading of Bitcoin's futures contracts has taken place in the Chicago Mercantile Exchange and the Chicago Board Options Exchange in December 2017. Brito et al. (2014) address the emergence of derivatives in the context of Bitcoin, with a focus on the regulatory aspects relating to it. In a more recent study, Corbet et al. (2018) examined the introduction of Bitcoin futures, and found that these are not effective hedging instruments, given that spot volatility has increased following the introduction of these contracts.

5 Economic Policy and Market Uncertainties

An important and intuitive observation noted by Ji et al. (2019) is that the role of a cryptocurrency as a transmitter or a receiver of a shock alternates, and the significance of the returns' connectedness and volatility spillovers changes throughout time. Many external factors, such as economic policy uncertainty (EPU), stock market uncertainty and the prices of other securities and commodities could affect the return of, and the dynamics of the connectedness among cryptocurrencies, and between cryptocurrencies and other conventional assets. Taking into consideration such aspects is crucial for investors when deciding their investment and risk management strategies depending on the economic and policy-related conditions.

There is evidence in the existing literature on the negative correlation between EPU and stock returns (Chiang 2019), and on the negative relation between EPU and stock prices (Kang and Ratti 2014; Antonakakis et al. 2013). Demir et al. (2018) found that the US EPU index can be used to predict Bitcoin returns, and is negatively correlated with these returns; therefore Bitcoin can be used as a hedge against EPU. They argue that, in a state of high economic policy uncertainty, investors tend to have a lower level of trust in the global financial systems and conventional currencies, and therefore they become more attracted to invest in Bitcoin. Matkovskyy and Jalan (2019) analyzed the effects of EPU on the interdependence between Bitcoin and traditional financial markets. Their study considered both the return- and volatility-related effects. They used five stock market indices (NASDAQ100, S&P500, Euronext100, FTSE100 and NIKKEI225), and measured EPU based on economic policy, monetary policy, financial regulation, taxation policy, and the news-based policy uncertainty index for the U.S., U.K., Europe and Japan. Their results show that, the connectedness between Bitcoin and traditional financial markets in terms of volatility is higher than their connectedness in terms of returns, and that EPU shocks have a negative impact on the interdependence between Bitcoin and traditional financial markets. The findings also provide support for the role of Bitcoin as a hedge against US economic uncertainty shocks.

Fang et al. (2019) examined the effects of global economic policy uncertainty (GEPU) on the long-run volatilities of Bitcoin, global equities, commodities, and bonds, using the GARCH-MIDAS model and its extension, the DCC-MIDAS model. Their findings revealed that the global economic uncertainty has a significant effect on the long-term volatility of Bitcoin, equities, and commodities, a negative significant impact on the Bitcoin-bonds correlation, and a positive impact on the Bitcoin-equities and Bitcoin-commodities correlations. Such findings suggest that Bitcoin can be used as a hedge against bonds during periods of high GEPU, and against equities and commodities during periods of low GEPU. Through further investigation, they found a weak effect of the state of economic uncertainty on the hedging abilities of Bitcoin. Bitcoin hedging abilities against equities and bonds have increased only slightly after accounting for the effect of the economic policy uncertainty. This led them to conclude that Bitcoin's hedging abilities are not only conditional on the strength of the GEPU, but also on how the other markets are related to it, and that the GEPU has a stronger effect on the volatility of global stock and bond indices than on Bitcoin.

Using quantile and quantile-on-quantile regressions, Bouri et al. (2017b) investigated the hedging properties of Bitcoin against global uncertainty, measured by the first principal component of the volatility indices (VIXs) of 14 developed and developing stock markets. The VIX is an indicator of market uncertainty as it reflects market sentiment and investor expectations. Higher VIX values signal higher market uncertainty. The results obtained highlight the importance of exploring the different investment horizons of Bitcoin returns, rather than just studying the entire conditional distribution of Bitcoin returns or just the conditional mean. For short-term horizons, Bitcoin is found to display hedging properties against global uncertainty only when the market is performing well (in a bull market context). In addition to the market conditions (bull or bear markets), the degree of uncertainty (whether there is a high or a low level of uncertainty) is also found to affect the hedging abilities of Bitcoin against uncertainty. Bitcoin was found to act as a hedge against uncertainty for shorter investment horizons at extreme ends of Bitcoin returns and uncertainty (when uncertainty is too high or too low). In a more recent study, Bouri et al. (2018) examined the quantile conditional dependence and causality between Bitcoin returns and the Global Financial Stress Index (GFSI). Using copula-based approach, they found that Bitcoin can act as a safe-haven against global financial stress.

To explore the prediction power of the daily EPU index on the daily Bitcoin returns, Demir et al. (2018) used the Bayesian Graphical Structural Vector Autoregressive model, the Ordinary Least Squares and the quantile-on-quantile Regression. Their findings revealed that the EPU has a predictive power on Bitcoin returns, and that Bitcoin returns are negatively associated with the EPU; an increase in the EPU results in a decrease in Bitcoin returns. However, this relation does not hold at the extreme ends of Bitcoin returns and uncertainty. At both the lower and higher quantiles of Bitcoin returns and EPU the effect becomes positive and significant. This finding is in line with the findings of Bouri et al. (2017b) that support the hedging abilities of Bitcoin against uncertainty during times of bull-market, and its diversification abilities during times of bear-market. Given the potential effect of a high level of policy uncertainty on investors' trust towards the economy and conventional currencies, one could justify the consequences of such conditions on cryptocurrencies' returns; under such conditions investors find Bitcoins more attractive.

Wang et al. (2018) used the US EPU index, the VIX and the equity market uncertainty index as proxies for EPU to investigate the risk spillover effect from EPU to Bitcoin. In terms of proxying for EPU, this study provides more comprehensive measures compared to Demir et al. (2018) and Bouri et al. (2017b) who have, each, used only one proxy for EPU (US EPU index or the VIX). Using a multivariate quantile model (MVQM) and the Granger causality risk test, on daily and weekly Bitcoin and EPU data, they found that the EPU has, in general, a negligible risk spillover effect on Bitcoin. These results, despite not being aligned with Demir et al.'s (2018) and Bouri et al.'s (2017b) findings, they supported the researchers' initial hypothesis; Wang et al. argue that, given the independence of Bitcoin with regards to the economic and financial system, one would expect the effect of EPU shocks on Bitcoin to be negligible, or non-existent. This weak spillover effect enables Bitcoin to act as a safe-haven or a diversifier when there is a high level of economic policy uncertainty.

6 Cryptocurrencies and Gold

A widespread view on cryptocurrencies' hedging abilities involves outlining its similarities with gold, in terms of scarcity of supply, high price volatility, existence of a finite supply, decentralization and lack of government control. A considerable literature thread found empirical evidence on the similarities between gold and cryptocurrencies due to their positive role in portfolio and risk management (i.e. Dyhrberg 2016a, b; Tully and Lucey 2007; Baur 2012). Cryptocurrencies have even been referred to as the new gold (Klein et al. 2018) or digital gold (Popper 2015). On the other hand, gold exhibits significant differences compared to Bitcoin; such as "tangibility, long history, intrinsic value, low volatility, and usage in the production process" (Al-Khazali et al. 2018).

In order to explore the similarities between Bitcoin and gold in terms of hedging abilities, Dyhrberg (2016b) examined the hedging capabilities of Bitcoin using a research approach (asymmetric GARCH methodology) and explanatory variables that are similar to the ones used in studying the hedging abilities of gold. Such an approach allows for a better comparison of the research findings on the two assets. The results revealed that Bitcoin is, on average, uncorrelated with the assets in the FTSE Index, so Bitcoin returns are not affected by changes in the stock market. This observation illustrates the role that Bitcoin can play in terms of reducing the market risk assumed by investors, and are in line with the findings on the hedging abilities of gold (Bauer and Lucey 2010). However, the hedging abilities of Bitcoin against the dollar appeared to be shorter lived than those of gold. She concluded that Bitcoin has a significant role in portfolio and risk management alongside gold.

In another study, Dyhrberg (2016a) questions whether Bitcoin is more similar to gold (as a store of value asset) or to the US dollar (as a medium for exchange). She identified similarities between Bitcoin and both gold and the US dollar. Bitcoin provides similar risk-management capabilities as gold, given their similarities in terms of their response to exchange rates' changes and large volatility persistence, and they both react symmetrically to good and bad news. She also found that Bitcoin reacts significantly to the US federal funds rate, which points to its role as a currency. An appreciation in the US dollar due to an increase in the federal funds rate, would lead to an increase in online purchases, and consequently to a higher demand, and improved returns, for Bitcoin. In conclusion, she argues that Bitcoin is a hybrid between the gold and the dollar and that it displays hedging capabilities of value in risk and portfolio management.

Gold prices are found to have a significant negative effect on Bitcoin's returns spillovers, which could be explained by the similarities between Bitcoin and gold in terms of their hedging abilities; when the gold price increases, the demand for Bitcoin decreases which would weaken its return spillover effect (Ji et al. 2019). Interestingly, Klein et al. (2018), in their paper entitled "Bitcoin Is Not the New Gold: A Comparison of Volatility, Correlation, and Portfolio Performance" challenge the mainstream view referring to cryptocurrencies as the new gold. They claim that "the two assets could barely be more different" (Klein et al. 2018) as they exhibit fundamentally different properties as assets and are differently linked to equity markets. The results showed differences in the conditional variance structures of the two assets, and in their correlations behaviors, especially in times of market distress. By evaluating time-varying conditional correlations, using a BEKK-GARCH model, they found that Bitcoin moves in the same direction as the stock markets during down times, which is completely different than the way gold behaves in downward markets. Their findings also contrast with a considerable existing literature thread that highlights the hedging abilities of Bitcoin; Klein et al. concluded that Bitcoin has unstable hedging properties and is not a safe-haven.

In the same vein, Al-Khazali et al. (2018) explored the impact of macro-economic news surprises on the returns and volatilities of gold and Bitcoin, based on a dataset originating from the US, Canada, the Euro Area, UK, and Japan. Using the GARCH methodology, they found gold and Bitcoin to display asymmetric reactions to these news; handing support to the evidence on the difference between the two. Whereas the gold's response exhibited its safe-haven properties, Bitcoin, in general, behaved differently. The noted negative co-movement of gold prices with macro-economic news highlights its safe-haven capabilities, which has also been documented in previous studies (Elder et al. 2012; Bauer and Lucey 2010).

References

Al-Khazali O, Bouri E, Roubaud D (2018) The impact of positive and negative macroeconomic news surprises: gold versus Bitcoin. Econ Bull 38(1):373–382

Antonakakis N, Chatziantoniou I, Filis G (2013) Dynamic co-movements of stock-market returns, implied volatility and policy uncertainty. Econ Lett 120:87–92

Antonakakis N, Chatziantoniou I, Gabauer D (2019) Cryptocurrency market contagion: market uncertainty, market complexity, and dynamic portfolios. J Int Financ Mark Inst Money 61:37–51

Bauer DG, Lucey M (2010) Is gold a hedge or a safe haven? An analysis of stocks, bonds and gold. Financ Rev 45(2):217–229

Baur D (2012) Asymmetric volatility in the gold market. J Altern Invest 14(4):26–38

Baur DG, McDermott TK (2010) Is gold a safe haven? International evidence. J Bank Financ 34 (8):1886–1898

Baur DG, Lee AD, Hong K (2015) Bitcoin: currency or investment? Available at SSRN: 2561183

Baur DG, Hong K, Lee AD (2018) Bitcoin: medium of exchange or speculative assets? J Int Financ Mark Inst Money 54:177–189

Beneki C, Koulis A, Kyriazis NA, Papadamou S (2019) Investigating volatility transmission and hedging properties between Bitcoin and Ethereum. Res Int Bus Financ 48:219–227

Berentsen A, Schar F (2018) A short introduction to the world of cryptocurrencies. Fed Reserve Bank St. Louis Rev 100:1), 1–1),16

Boako G, Tiwari AK, Roubaud D (2019) Vine copula-based dependence and portfolio value-at-risk analysis of the cryptocurrency market. Int Econ 158:77–90

Borri N (2019) Conditional tail-risk in cryptocurrency markets. J Empir Financ 50:1–19

Bouri E, Azzi G, Dyhrberg AH (2017a) On the return-volatility relationship in the Bitcoin market around the price crash of 2013. Econ Open-Assess E-J 11:1–16

Bouri E, Gupta R, Tiwari AK, Roubaud D (2017b) Does Bitcoin hedge global uncertainty? Evidence from wavelet-based quantile-in-quantile regressions. Financ Res Lett 23:87–95

Bouri E, Molnár P, Azzi G, Roubaud D, Hagfors LI (2017c) On the hedge and safe haven properties of Bitcoin: is it really more than a diversifier? Financ Res Lett 20:192–198

Bouri E, Gupta R, Lau CKM, Roubaud D, Wang S (2018) Bitcoin and global financial stress: a copula-based approach to dependence and causality in the quantiles. Q Rev Econ Financ 69:297–307

Brauneis A, Mestel R (2018) Price discovery of cryptocurrencies: Bitcoin and beyond. Econ Lett 165:58–61

Brière M, Oosterlinck K, Szafarz A (2015) Virtual currency, tangible return: portfolio diversification with Bitcoin. J Asset Manage 16(6):365–373

Brito J, Shadab H, Castillo A (2014) Bitcoin financial regulation: securities, derivatives, prediction markets, and gambling. Columbia Sci Technol Law Rev 144:2014–2015

Canh NP, Wongchoti U, Thanh SD, Thong NT (2019) Systematic risk in cryptocurrency market: evidence from DCC-MGARCH model. Financ Res Lett 29:90–100

Caporale GM, Gil-Alana L, Plastun A (2018) Persistence in the cryptocurrency market. Res Int Bus Financ 46:141–148

Chan WH, Le M, Wu YW (2019) Holding Bitcoin longer: the dynamic hedging abilities of Bitcoin. Q Rev Econ Financ 71:107–113

Charfeddine L, Maouchi Y (2018) Are shocks on the returns and volatility of cryptocurrencies really persistent? Financ Res Lett 28:423–430

Cheah ET, Fry J (2015) Speculative bubbles in Bitcoin markets? An empirical investigation into the fundamental value of Bitcoin. Econ Lett 130:32–36

Chen YW, Vivek KP (2014) The value of Bitcoin in enhancing the efficiency of an investor's portfolio. J Financ Plan 27(9):44–52

Chiang CT (2019) Economic policy uncertainty, risk and stock returns: evidence from G7 stock markets. Financ Res Lett 29:41–49

Ciaian P, Rajcaniova M, Kancs DA (2018) Virtual relationships: short- and long-run evidence from Bitcoin and altcoin markets. J Int Financ Mark Inst Money 52:173–195

Corbet S, McHugh G, Meegan A (2017) The influence of central bank monetary policy announcements on cryptocurrency return volatility. Invest Manage Financ Innov 14(4):60–72

Corbet S, Meegan A, Larkin C, Lucey B, Yarovaya L (2018) Exploring the dynamic relationships between cryptocurrencies and other financial assets. Econ Lett 165:28–34

Demir E, Gozgor G, Lau CKM, Vigne SA (2018) Does economic policy uncertainty predict the Bitcoin returns? An empirical investigation. Financ Res Lett 26:145–149

Demirer R, Kutan AM (2006) Does herding behavior exist in Chinese stock markets? J Int Financ Mark Inst Money 16(2):123–142

Diebold FX, Yilmaz K (2014) On the network topology of variance decompositions: measuring the connectedness of financial firms. J Econ 182:119–134

Dyhrberg AH (2016a) Bitcoin, gold and the dollar – a GARCH volatility analysis. Financ Res Lett 16:85–92

Dyhrberg AH (2016b) Hedging capabilities of Bitcoin. Is it the virtual gold? Financ Res Lett 16:139–144

Eisl A, Gasser SM, Weinmayer K (2015) Caveat Emptor: does Bitcoin improve portfolio diversification? SSRN Electronic Journal

Elder J, Miao H, Ramchander S (2012) Impact of macroeconomic news on metal futures. J Bank Financ 36(1):51–65

Fama E (1970) Efficient capital markets: a review of theory and empirical work. J Financ 25 (2):383–417

Fang L, Bouri E, Gupta R, Roubaud D (2019) Does global economic uncertainty matter for the volatility and hedging effectiveness of Bitcoin? Int Rev Financ Anal 61:29–36

Feng W, Wang Y, Zhang Z (2018) Can cryptocurrencies be a safe haven: a tail risk perspective analysis. Appl Econ 54:4745–4762

Gkillas K, Katsiampa P (2018) An application of extreme value theory to cryptocurrencies. Econ Lett 164:109–111

Guesmi K, Saadi S, Abid I, Ftiti Z (2018) Portfolio diversification with virtual currency: evidence from Bitcoin. Int Rev Financ Anal 63:431–437
Halaburda H, Gandal N (2014) Competition in the cryptocurrency market. NET Institute Working Paper No. 14-17
Ji Q, Bouri E, Mau CKM, Roubaud D (2019) Dynamic connectedness and integration in cryptocurrency markets. Int Rev Financ Anal 63:257–272
Kang W, Ratti RA (2014) Oil shocks, policy uncertainty and stock-market returns. J Int Financ Mark Inst Money 26:305–318
Klein T, Thu HP, Walther T (2018) Bitcoin is not the new gold – a comparison of volatility, correlation, and portfolio performance. Int Rev Financ Anal 59:105–116
Matkovskyy R, Jalan A (2019) Effects of economic policy uncertainty shocks on the interdependence between cryptocurrency and traditional financial markets. Cryptocurrency Research Conference 2019
Molnár P, Vagstad K, Valstad OCA (2015) A bit risky? A comparison between Bitcoin and other assets using an intraday value at risk approach. Working Paper
Nadarajah S, Chu J (2017) On the inefficiency of Bitcoin. Econ Lett 150:6–9
Pal D, Mitra SK (2019) Hedging Bitcoin with other financial assets. Financ Res Lett 30:30–36
Polasik M, Piotrowska AI, Wisniewski TP, Kotkowski R, Lightfoot G (2015) Price fluctuations and the use of Bitcoin: an empirical inquiry. Int J Electron Commerce 20(1):9–49
Popper N (2015) Digital gold: the untold story of Bitcoin. Penguin, London
Selmi R, Mensi W, Hammoudeh S, Bouoiyour J (2018) Is Bitcoin a hedge, a safe haven or a diversifier for oil price movements? A comparison with gold. Energy Econ 74:787–801
Sun X, Liu M, Sima Z (2018) A novel cryptocurrency price trend forecasting model based on LightGBM. Financ Res Lett
Tully E, Lucey BM (2007) A power GARCH examination of the gold market. Res Int Bus Financ 21:316–325
Urquhart A (2016) The inefficiency of Bitcoin. Econ Lett 148:80–82
Urquhart A, Zhang H (2018) Is Bitcoin a hedge or safe-haven for currencies? An intraday analysis. Working Paper 3114108. SSRN
Vidal-Tomás D, Ibañez A (2018) Semi-strong efficiency of Bitcoin. Financ Resh Lett 27:259–265
Wang GJ, Xie C, Wen D, Zhao L (2018) When Bitcoin meets economic policy uncertainty (EPU): measuring risk spillover effect from EPU to Bitcoin. Financ Res Lett
Wei WC (2018) Liquidity and market efficiency in cryptocurrencies. Econ Lett 168:21–24
Wong WS, Saerbeck D, Delgado Silva D (2018) Cryptocurrency: a new investment opportunity? An investigation of the hedging capability of cryptocurrencies and their influence on stock, Bond and Gold Portfolios. SSRN Electronic Journal
Yermack D (2013) Is Bitcoin a real currency? An economic appraisal. National Bureau of Economic Research. https://www.nber.org/papers/w19747
Yi S, Xu Z, Wang GJ (2018) Volatility connectedness in the cryptocurrency market: is Bitcoin a dominant cryptocurrency? Int Rev Financ Anal 60:98–114

Index

A

Accountability, 167, 171, 172, 175
Anti-hacking measurement, 82
Application programming interface (API), 59
Application-specific integrated circuit (ASIC), 58
Artificial Intelligence (AI), 91
Asset classes, 1, 182, 183
Auditing process, 109
Australian Transaction Reports and Analysis Centre (AUSTRAC), 30
Automated-teller machine (ATM), 82

B

Basic Attention Token (BAT), 93
BFGMiner, 59
Bitcoin, 1, 4, 6–10, 18, 20, 29, 37, 40, 92, 115, 183
  coffee bonding theory, 69–71
  cryptocurrency, 69
  direct implications, 69
  double spending problem, 69
  mining rules, 34
  network, 61
  price, 35
  rules, 36
  tax case study
    involuntary noncompliance, 72, 73
    legitimacy, 71
    tax regulators, 71
    virtual currencies, 71
    voluntary noncompliance, 73
  virtual currency and comprehensive regulation, 69
Bitcoin Energy Consumption Index, 60
Bitcoin mining, 34, 38, 55, 64
Bitcoin system, 56
Blockchain
  Brave Browser, 93
  coin classification framework
    asset backed token, 102
    cryptocurrency, 101
    debt token, 102
    equity token, 101
    hybrid token, 103
    international movement, 103
    investment composite and indexes, 102
    security token, 101
    stablecoin, 102, 103
    utility tokens, 101
  cryptocurrencies, 92
  cryptoKitties, 92, 93
  crypto-recession/crypto-winter, 97
  digital assets, 94
  digital values, 92
  Hosho, 96
  ICO, 94 (*see also* Initial coin offering (ICO))
  institutional investors, 96
  multi-purpose technology, 91
  opportunity, 96
  regulatory compliance, 96
  remote supporters/community, 96
  smart contracts, 96
  team and communities, 96
  transformation, 91
  uPort, 93, 94
Blockchain technologies, 77–79, 85, 87
Business ethics, 43

S. Goutte et al. (eds.), *Cryptofinance and Mechanisms of Exchange*, Contributions to Management Science, https://doi.org/10.1007/978-3-030-30738-7

C

Canadian Parliament, 47
Central banks, 44
Central processing unit (CPU), 57
CGminer, 59
Coffee bonding theory, 70, 71
Commodities and Futures Trading Commission (CFTC), 69, 85, 86, 172–175
Companies' Creditors Arrangement Act, 86
Computer system, 56
Computing technology, 36
Consensus mechanisms, 5
Consensys Formation, 94
Cooling, 60
Corporate governance, 132
  and security tokens, 133
  and utility tokens, 133
Credit card transactions, 44
Credit Protocol system, 102
Crohn's disease, 82
Crowdfunding, 146
Cryptoanarchism, 87
Crypto CAPM, 137
Cryptocurrencies, 39, 43, 46, 92
  accountability, 167
  ASIC, 65
  asset class, 2, 3, 6
  autonomous, 77
  bitcoin, 14, 126, 129
  blockchain architecture, 2, 4, 5
  correlation analyses, 12
  correlation and liquidity analyses, 9, 10
  CPUs and GPUs, 65
  CVaR, 9
  cybercrime, 39
  DApp, 6, 7
  decentralization, 77
  development, 5
  direct human intervention and participation, 78
  domain of, 87
  electricity price, 64
  encryption protocol, 3
  exchanges
    matching platforms, 81
    Mt. Gox, 81, 82
    proliferation, 81
    Quadriga CX, 82
    sale and purchase, 81
  exogenous, 87
  ex-post optimal weightings, 15
  FAANG stock pairs, 12
  fake and fraudulent, 170
  and gold, 189, 190
  hard forks, 80
  hedging cryptocurrency investments, 183–186
  human agency, 78
  ICO, 84, 85
  illegal activities, 39
  illegal financial transactions, 38
  immutability, 77
  infractions, 78
  international regulatory responses, 171
  internet, 38
  investability, 7–9
  investors and network participants, 78
  lack of third-party verification requirements, 78
  legitimacy, 6
  liquidity, 8, 13
  literature, 2, 9
  long-term viability, 78
  LULD, 14
  market capitalization, 185
  market efficiency, 180, 181
  market liquidity, 2
  market stability, 13
  mechanisms, 4
  methodology, 12
  mining puzzles, 65
  monetary systems, traditional finance, 77
  MWCBs, 13
  popular discourse, 78
  popularity, 169
  portfolio performance, 2
  PoW mechanism, 4, 5
  principle assertion, 88
  promise de trustlessness, 79
  protocols, 3
  Quadriga CX case, 86
  recourse, traditional structures, 85
  regulation and oversight, 88
  regulatory gap, ICOs, 169
  risk mitigation, 181–183
  SEC and CFTC, 85, 86
  socioeconomic realm, 78
  suitability and performance, 6
  Swiss initiative, 6
  terrorism, 41
  traditional asset classes, 2, 11, 12
  trustlessness, 78, 87
  USDT, 10
  volatile prices, 64, 167
  zero bits, 4
Cryptocurrency mining, 52, 60

blockchain, 52
computational power, 53
costs and factors, 53
CPU, 57
cryptocurrency protocol, 51
miners, 52
mining contracts, 54
POW, 53
SHA-256 transforms, 53
solo mining, 53
transaction data, 52
transactions, 52
Cryptocurrency mining hardware, 58
Cryptocurrency Tax Fairness Act, 72
Crypto-currency tokens, 127
Cryptoexchanges
consequences, pumps and dumps, 100
cryptocurrencies, 97
ICO and traditional stock market, 98, 99
market maker, 97
pump and dump online community, 99, 100
traditional payment processor, 98
types of, 98
Cryptographic tokens, 18
CryptoKitties, 92, 93
Crypto portfolios, 22
Crypto tokens, 145, 155
crypto-currency, 127
security tokens, 126
token offerings *vs.* IPOs, 127–128
types, 126
utility tokens, 126
CryptoValley, 6
Current utility value (CUV), 135, 136
Cyberattacks, 41
Cybercrime, 39
Cybersecurity, 82

D

Dark ethical markets, 47
Decentralized Applications (dApps), 5, 92
Decentralized Autonomous Organization (DAO), 5, 77, 80, 132, 133
Digiconomist method, 33
Digital age, 44
Digital currencies, 42
Digital currency portfolios, 45
Digital wealth creation, 61
Direct human intervention, 80
Discounted cash flow (DCF) analysis, 137
Discounted expected utility value (DEUV), 135, 136
Distributed organization models, 133
Diversifier, 179
Dynamic approach, 21
Dynamic conditional correlation GARCH (DCC-GARCH) method, 182

E

EasyMiner, 59
Economic-based approach, 62
Economic policy uncertainty (EPU), 186, 187
Economic pressures, 83
Economic turbulences, 1
Efficient market hypothesis, 180
Energy consumption, 64
EPU index, 188
Equity curves, 23
Ether (ETH), 115
Ethereum, 92, 126, 181, 184, 185
Ethereum community roadmap, 37
Ethereum Request for Comment 20 (ERC20), 115
Ethereum systems, 32
Euro backed stablecoin (EURT), 102
European Central Bank, 47

F

Fair trade, 47
"Fear of missing out" (FOMO), 171
Field programmable gate array (FPGA), 58
Financial Action Task Force (FATF) report, 40
Financial Crimes Enforcement Network (FinCEN), 69
Financial intermediation, 44
Financial market data, 11
Financial technology sector, 47
Financing illegal activities, 39–41

G

Gemini dollar (GUSD), 102
Generalized Orthogonal GARCH (GO-GARCH) model, 186
Global economic policy uncertainty (GEPU), 187
Gold, 183, 184, 189, 190
Graphic user interface (GUI), 59
Graphical Processing Units (GPU), 57

H

Hard cap, 116, 152
Hardware and software requirements, 59
Hash rate distribution, 55
Hosted mining, 54
Howey Test, 127

I

ICO regulations, 166, 167, 171, 174
ICO sale
  average duration, 160
  exchange listing, 161
  investor protection, 160
  market liquidity, 160
  post-ICO, 160
  presale, 158, 159
  token demand, 160
Individual cryptocurrency liquidity, 19
Information Reporting Advisory Committee, 71
Initial Coin Offering (ICO), 78, 84, 85, 113
  academic interest, 167–169
  advantage, 114
  behavioral biases, 140
  benefits, to entrepreneurs, 129–130
  conceivable advantages, 168
  cumulative analysis, 170
  drawbacks, 131
  form of investment, 114
  full-cycle ICO process, 146
  geography, 147
  global phenomenon, 131
  history, 114, 115
  long-term returns and performance, 139
  mainstream, 170
  market, 146
  online realm, 166
  post-ICO stage, 147
  pre-ICO stage, 147
  regulatory responses, 171–174
  security tokens, 145
  selection of qualitative parameters, 114
  set of data, 113
  stages, 147
  technical elements, 131
  token sale/crowdsale, 166
  traditional capital market, 114
  underpricing and first-day returns, 138, 139
  utility tokens, 145
Initial Exchange Offering (IEO)
  contra-arguments, STOs, 109
  crypto exchange, 108
  launchpads, 108
  team and projects, 108
  win for the exchange, 108
  win for the investors, 108
  win for the start-up, 108
Initial Public Offerings (IPOs), 114
Innovative technology, 46
Innovative VAR methodologies, 184
Internal Revenue Service (IRS), 69
Internet, 40
Internet of Things (IoT), 91
Investment tokens, 126

K

Know your customer (KYC) policies, 31, 154, 155
Kolmogorov-Smirnov test, 12

L

Leased hashing power, 54
Limit up-limit down levels (LULD), 13
Limit-up limit-down triggering, 20
Liquidity, 13, 18
Listing, 161

M

Machine learning models, 79
Macroeconomic level, 44
Macroeconomic risk factors, 65
Market activity, 83
Market efficiency, 180, 181
Market stability, 13
Market wide circuit breaks (MWCB), 2, 13, 20
Mean-variance approach, 9
Minimum viable product (MVP), 104
Mining, 72
Mining competition, 35, 37
Mining energy efficiency, 36
Mining equipment, 35
Mining pools, 55, 56
Mining process, 63
Money laundering, 42
Motherboard, 57
MultiMiner, 59
Multivariate quantile model (MVQM), 188

N

Network effects, 129, 130
Network value-to-transaction ratio (NVT), 136
New York Stock Exchange, 97

P

Planning stages, ICOs
  ICO ratings and online forums, 157–158
  legal aspects
    KYC and AML, 154, 155
    sales restrictions, 153, 154

social media, 156, 157
technology and source code, 155
token design and tokenomics
currency accepted, 153
hard cap, 152
lock-up mechanism, 153
soft cap, 151, 152
token supply, 152
token type, 150, 151
white paper, 147
legal aspects, 148
management team and advisers, 149, 150
proceeds and ICO success, disclosure of use, 149
quality and informativeness, 148
technical aspects, 149
Poloniex supplies quotes against the Tether (USDT), 10
Portfolio diversification, 181–183
Portfolio optimization, 14, 21–23
Portfolio optimization parametric and non-parametric tests, 16
Portfolio optimization simulations, 11
Price dynamics, 185
Privacy-preservation mechanism, 79
Proof of Burn (PoB), 4
Proof of Stake (PoS), 4
Proof of Work (PoW), 4, 52, 62
Protocol software, 80

Q
Quantity Theory of Money (QTM), 134, 135

R
Ransomware, 41
Regulatory authorities, 83
Risk mitigation, 181, 182

S
Safe-haven, 180, 181
Scrypt algorithm, 66
Securities and Exchange Commission (SEC), 69, 85, 86, 172–174
Securities Exchange Commission (SEC), 101
Security Token Offerings (STO), 145
advantages and disadvantages, 106, 107
crypto market, 103
decentralized project, 103
investors and regulators, 104
process of issuing, 104
Regulation D (Reg D), 104, 105
regulatory frameworks, 104
steps into launching, 105, 106
type of investors, 104
Security tokens, 126
advantages, 126
and corporate governance, 133
characteristics, 126
valuation, 137
Self-Regulatory Organizations (SRO), 13
Self-Sovereign Identity, 94
Small-and large-scale investors, 85
Social media, 156, 157
Soft cap, 151, 152
Speculative bubbles, 140

T
Tax evasion, 71, 73, 74
Terrorist financing, 40
Thermal design power (TDP), 61
Time stamping and witnessed digital signatures, 3
Token fungibility *vs.* non-fungibility, 92
Token issuing, 130
Token offerings
Howey Test, 127
vs. IPOs, 127, 128
security, 125
and tokenomics, 127
utility, 125
Tokenomics, 96, 99
definition, 113, 128, 129
economic design, 113
economic value, 113
further research, 122
parameters, 113
capital collection, 120
creation, tokens, 118
currency acceptance, 119
distribution schedules, 117
hard cap, 120
investors, 117
KYC processes, 116
literature research, 116
literature review, 116
money collection, 118
private sales, 116, 120
public ICO price, 115
public sales, 116, 120
sale proceeds, 118
sale process, 116, 117
sale quota, 117, 120
sample findings and the literature research findings, 120
segmentation, 115
soft cap, 116, 120

Tokenomics (*cont.*)
standard currency, 119
standard issue price, 119
technical standard, 115, 119
token allocation, 117, 120
token type, 115, 119
tokenomics, 118
total supply, 120
venture, 115
set of data, ICOs, 113
token economics, 126
Token price, 132, 133, 135, 136
Token supply, 152
Toronto Stock Exchange, 97
Traditional asset classes, 16
Traditional payment systems, 62
Traditional Stock Market, 98, 99
Treasury Inspector General for Tax Administration (TIGTA), 72
Trust-based activity, 80

U
UK National Crime Agency's survey, 42
Underpricing, 138, 139
US Federal Bureau of Investigation (FBI), 83
Utility tokens, 126
and corporate governance, 133
description, 125
issuance, 127

V
Valuation, crypto tokens
crypto J-curve methodology, 135, 136
fundamental analysis, 134
NVT, 136
security token valuation, 137
token velocity methodology, 134, 135
traditional valuation methods, 134
comparables approach, 138
crypto CAPM, 137
DCF, 137
Venture capital companies (VCs), 116
Venture capitalists, 94
Virtual currencies, 63
AUSTRAC report, 30
bitcoin and currencies, 30
bitcoin and Ethereum, 32
cryptocurrencies, 30
cryptomining, 36
e-commerce, 31
electricity consumption, 34
electricity price, 35
energy consumption, 33
energy justice, 31
financial stability, 31
global warming, 32
green electricity, 36
market capitalization, 29
services offering, 31
stakeholders, 37
system parameters, 36–37
world energy production system, 32
Virtual environments, 43
Virtual hosted machines, 54
Virtual money laundering, 38
Visa and MasterCard systems, 44
Visa electronic payment system, 33
Vitalik Buterin, 94
Volatility indices (VIXs), 187

The manufacturer's authorised representative in the EU is Springer Nature Customer Service Centre GmbH, Europaplatz 3, 69115 Heidelberg, Germany. If you have any concerns regarding our products, please contact ProductSafety@springernature.com

Printed and bound by CPI Group (UK) Ltd, Croydon, CR0 4YY
22/09/2026
02221058-0006